Comprehensive
NANOBIOTECHNOLOGY

Comprehensive NANOBIOTECHNOLOGY

V. SREE KRISHNA

M.Sc. (Biological Sciences), Ph.D. (Plant Biotechnology),
Ph.D. (Marine Biotechnology) and M.B.A.

Principal
Osteen Degree College, Bangalore
Karnataka

PUBLISHING FOR ONE WORLD

NEW AGE INTERNATIONAL (P) LIMITED, PUBLISHERS

New Delhi • Bangalore • Chennai • Cochin • Guwahati • Hyderabad
Jalandhar • Kolkata • Lucknow • Mumbai • Ranchi
Visit us at **www.newagepublishers.com**

Published by New Age International (P) Ltd., Publishers

First Edition: 2011

Branches:

- No. 37/10, 8th Cross (Near Hanuman Temple), Azad Nagar, Chamrajpet, **Bangalore**-560 018. Tel.: (080) 26756823, Telefax: 26756820, E-mail: bangalore@newagepublishers.com
- 26, Damodaran Street, T. Nagar, **Chennai**-600 017. Tel.: (044) 24353401, Telefax: 24351463 E-mail: chennai@newagepublishers.com
- CC-39/1016, Carrier Station Road, Ernakulam South, **Cochin**-682 016. Tel.: (0484) 2377004, Telefax: 4051303. E-mail: cochin@newagepublishers.com
- Hemsen Complex, Mohd. Shah Road, Paltan Bazar, Near Starline Hotel, **Guwahati**-781 008. Tel.: (0361) 2513881. Telefax: 2543669, E-mail: guwahati@newagepublishers.com
- No. 105, 1st Floor, Madhiray Kaveri Tower, 3-2-19, Azam Jahi Road, Nimboliadda, **Hyderabad**-500 027. Tel.: (040) 24652456, Telefax: 24652457, E-mail:hyderabad@newagepublishers.com
- RDB Chambers (Formerly Lotus Cinema)106A, 1st Floor, S.N. Banerjee Road, **Kolkata**-700 014. Tel.: (033) 22273773, Telefax: 22275247, E-mail:kolkata@newagepublishers.com
- 16-A, Jopling Road, **Lucknow**-226 001. Tel.: (0522) 2209578, 4045297, Telefax: 2204098 E-mail: lucknow@newagepublishers.com
- 142C, Victor House, Ground Floor, N.M. Joshi Marg, Lower Parel, **Mumbai**-400 013. Tel.: (022) 24927869. Telefax: 24915415, E-mail: mumbai@newagepublishers.com
- 22, Golden House, Daryaganj, **New Delhi**-110 002. Tel.: (011) 23262370, 23262368, Telefax: 43551305. E-mail: sales@newagepublishers.com

---

ISBN: 978-81-224-3082-0

Rs. 125.00

C-10-06-4672

Printed in India at Ram Printograph, Delhi.
Typeset at Goswami Associates, New Delhi.

PUBLISHING FOR ONE WORLD
NEW AGE INTERNATIONAL (P) LIMITED, PUBLISHERS
4835/24, Ansari Road, Daryaganj, New Delhi-110002
Visit us at **www.newagepublishers.com**

Dedicated to
All my benefactors, teachers
and wellwishers, without their
cooperation I would not have
achieved the knowledge and capability

Preface

Nanotechnology has initiated with reference to study about very small sized things which can be measured by nanometers. Nanobiotechnology is an interdisciplinary area where research works are crossing over a very broad range of academic disciplines from Biology to Chemistry, Physics and Engineering.

The book **Comprehensive Nanobiotechnology** provides the readers with an introduction to current state and future prospects of this breakthrough technology. It offers an enticing look into a future engrossed in nanobiotechnological applications with the help of overviews on the enabling technologies being used currently in various biotechnology related industries.

Potential of nanobiotechnology to manipulate and programme matter with atomic precision has invited the attention of scientists to explore innumerable applications of nanobiotechnology to mankind.

The increasing importance of nanobiotechnology is an inspiration for the author by contributing his ideas, observations and findings for the benefit of students of M.Sc. (Biotechnology), B.E. (Biotechnology), research scholars, academicians and industrialists associated with it or its allied field.

V. Sree Krishna

Acknowledgement

I acknowledge my sincere thanks to all my wellwishers who have encouraged me to write this book. I extend my thanks to Sri Saumya Gupta (Managing Director), New Age International (P) Ltd., Publishers for accepting my proposal to write this book. I am thankful to Dr. R.C. Gupta (Academic Director), for his valuable suggestions. The author acknowledges his profound thanks to the editorial team of New Age International (P) Ltd., Publishers for showing keen interest in editing this book to this presentable level. I also extend my sincere gratitudes to Mr. Sudharshan (Marketing Manager) and all other staff of Bangalore branch for their utmost support in bringing this book out.

Last but not least I am especially thankful to my wife Mrs. S. Vijayasree and my son Mr. S. Gopalakrishna for having spared their invaluable time and patience in providing privacy, thus enabling me to author this book with devoted dedication.

V. Sree Krishna

Contents

Introduction

Nanotechnology is the creation and also utilisation of materials, devices, and systems through the control of matter on a nanometer-length scale; at the level of atoms, molecules, and supramolecular structures. In other words nanotechnology is the ability to works at these different levels, to generate larger structures with fundamentally new molecular organisation. The nanostructures made up of the smallest human made objects, and they exhibit novel physical, chemical and biological properties including phenomena.

The aim of nanotechnology is to learn to exploit these properties and efficiently manufacture and employ these structures.

Nanotechnology can be defined as the popular term for the construction and utilization of functional structures with at least one characteristic dimension in nanometers. When characteristic structural features are intermediate in extent between isolated smaller atoms and bulk materials in the range of about 10^{-9} to 10^{-7} meter, the objects often display physical dimensions which are totally different from those displayed by either bulk materials or atoms.

The properties of matter at the nanoscale are not predictable from those observed at larger scale. Changes of behaviour are not only caused by continuous modification of characteristics with diminishing size, but also emergence of totally new phenomenon, such as quantum size confinement, wave like transport, and predominant interfacial phenomenon. Once the ability of shape and feature size are achieved, it is possible to increase the material properties and device functions beyond the already established ones. It already exists in nature. For example if a human body is deconstructed into its basic ingredients it will be in huge quantity of oxygen, nitrogen, hydrogen, calcium and carbon besides little quantities of phosphorous, sulphur, magnesium, iron and around twenty other chemical elements in smaller quantities. However, total value is not significant at all. Nature transforms these in animate ingredients into self generating, self perpetuating, self replicating components like DNA, protein, enzymes, hormones etc., that ultimately transform into living organisms. It has an immense value because, the basic process involves an engineering at the molecular or nanometer dimension. Till now known stable nanostructures include various entities as nanotubes, proteins, DNA and single-electron transistors that operate at room temperature. Atomic clusters are metastable nanostructures which have the potential of fabricating new devices out of them including new novel properties.

NANOTECHNOLOGY AND VISION

In 1959, the great vision of Sir Richard Feynman stimulated people to fabricate materials and devices at the atomic as well as molecular levels. He has also opened that a new class of instrumentation would be needed to manipulate and measure the properties of these "nano"-small structures. Feynman envisioned instrument, has been invented in 1980s; which includes scanning tunneling microscope, near field microscopes and atomic force microscopes. These sophisticated instruments are also considered as the eyes and fingers needed for nanostructure manipulation and measurement.

Nobel Laureate Ronald Hoffmann has commented four decades back, "Nanotechnology is the way of ingeniously controlling the building of small and large structures, with intricate properties; it is a way of precise, controlled building, it is the way of the future, with incidentally environmental beginners built in by design". Formerly chief scientist of IBM visions has said "I believe nanoscience and nanotechnology will be central to the next epoch of the information age, and will be as revolutionary as science and technology, at the micron scale have been since early 70's". Horst Stormer has expressed his vision as "Nanotechnology has given us tools, to play with the ultimate toy box of nature— *i.e.,* atoms and molecules. Everything is made from it. The possibilities to create new things appear limitless."

Rational farbrication and integration of nanoscale materials and devices herald a revolutionary age for science and technology provided we can discover and fully utilize their underlying principles. Nanoscience offers a new paradigm of materials manufacture, by submicron scale assembly, to create, entities from the basic ingredients.

Neal Lane, adviser to the US President for Science and Technology, has commented on Nanotechnology as "a branch of Science and Engineering Technology that will most likely produce the breakthroughs of future."

Abdul Kalam, Honourable President of India has suggested about Nanotechnology that "It represents the limiting extension of Moore's law in electronics and sets the path for the classical devices to small devices, a step to quantum devices and fundamentally different architectures for processors." It paves the way to precisely control the arrangement of impurities and defects with respect to each other in materials science with the opportunity of new generation of advanced composites. These are the visions of the personalities at the forefront of Science and Technology today. This envisages "Nano" a theme in almost all established disciplines. A significant assurance from the 20th century is, that predictions are expected be established particularly in this technology for several decades in the future, often fall far short of what is exactly accomplished, because one foresees only evolutionary changes, while scientific and technological revolutions are predicted to be very promissing.

The nanometer dimensions are structurally different as it is in the quantum length scale. The properties of microstructures, the basis for technologies that support microelectronics, are explained largely by the physics of macroscopic systems. But nanostructures, however, are natural home for engineered quantum effects. It can be viewed as a unique state of matter encompassing a promise for new and potentially useful products; for example as key components in information technology and devices which have unprecedented functions.

Added to this due to their very tiny size they possess a very high padring density that is useful for high speed information processing and higher volumetric capacity thus, enabling information storage.

Reasons to Adopt Nanotechnology

There are several reasons why an object of nanometer dimension differs; including that, it shows certain novel properties. The following illustrations can justify the situation.

Single electron tunneling: Being very small, the nanoparticles possess a high self-capacitance. Putting in an array, the capacitance will become very large because of inter-particle charge transfer and then the tunneling of electrons through the very potential sustaining barrier.

Quantum size effect (QSE): If the size of a metallic object is less than the typical de Broglie wavelength of electrons, then the classical picture of electrons trapped within hard wall boundaries is realistic. The conduction band is then supposed to have discrete level structure. At low temperature, one might be able to view this, and also, the I-V characteristics of this is not expected to be conventional under Ohmic functionality. Rather, it can be proved that for singly charged gold particle facilitating to see this effect, the typical size has to be less than 2.5 nanometer in diameter. This is referred to as quantum size effect.

Catalysis: Aggregation of very small particles always should have a higher surface area compared to their volume. This fact is accepted because most of the phenomenon happens on surfaces. For example, catalysis, photosynthesis, ice formation etc., occur on the surfaces.

Another fact is that, the nanoparticles do not necessarily have the same basic structure as its bulk. For example, gold in bulk has an fcc structure but nanoscale gold particles of 3–5 nanometer diameter have icosahedral structure that have more amount of kinds and edges than a cube. This is a very important issue for catalysis because it is now known that catalysis primarily happens on the steps and edges of the surfaces where the atoms are comparatively loosely bounded. Rather it is already established that the nanoscale gold particles have specific catalytic properties. For example,

(*i*) Oxidation of CO at 70°C temperature.

(*ii*) In partial oxidation reactions like propylene to propylene oxide at 50°C very high selectivity is exhibited.

(*iii*) Reduction of nitric oxide with H_2 using alumina-supported gold nanoparticles at almost room temperature.

These examples express the tremendous potential in introducing new chemical technology leading to revolutionary industrial production of chemicals. Nanoscale gold particles supported on Fe_2O_3 have recently been commercialized by National Research Institute Osaka in Japan as the 'deodorisers' in bathrooms. National Laboratory of Los Alamos has recently published the improved activity of hydrogen storage, which is considered as an enabled technology for automobile fuel cell applications. Several such examples can be highlighted expressing the advantages, that the mankind is to benefit through combining nanotechnology with available today's technology.

Macroscopic quantum tunneling: One ferromagnetic particle, which is smaller than the size of a single domain involves typically 10^3 to 10^4 spins. The energy required to this particle depends on the orientation of the magnetic moment, that again depends on the anisotropy of the particle for example shape anisotropy and magnetic anisotropy. If there is no applied field, the energy is an even function of magnetic moment *i.e.,* any local minimum is degenerated with respect to a transformation from M to – M. But at an applied field there will be several local minima with reference to one global minimum and then the question of tunneling of spins from metastable states come in a transition from M to – M. However, at high temperature it is very difficult to see this phenomenon because the spin rotation is in unison, enabling a coupling of moment with rotation. This phenomenon is termed as the quantum tunneling of spins.

Plastics: In this modern age plastics have become all pervading. But their stability is a serious question, because of the hazard of fire, and also for concomitant smoke and combustion products. It has been observed that consolidated nanocomposites by the addition of well-dispersed inorganic nanoparticles can reduce this characteristic of plastics considerably. Because the heat releasing rates of thermoset polymer based materials are reduced by 60% when intercalated with silicate clays in a proportion of 2–6% of weight. This envisages in self extinguishing characteristics.

Ceramics: Titania and alumina like nanophase ceramics made out of consolidation of nanoparticles, have proved to be readily formable into small samples, and their mechanical behaviour indicates a significant degree of difficulty under compression. This behaviour has been established through grain boundary sliding over one another under stress, without breaking the bounds, across the boundaries between grains as a result of diffusion repairing events occuring at incipient crackes.

Hard materials: Carbide nanocomposites with typical 30–40 nanometer particle size are found to have considerably increased hardness, wear resistance and fracture toughness compared to their conventional grain size counterparts. This will lead to a large commercial influence in future for machine tools, mega tools, wear parts and microtwist drill bits etc.

Coatings: Coatings with grain size smaller than 100 nanometer are thermally inherently stable and efficiently block dislocation movement giving way to extreme hardness, and in some cases four times thicker than those made with conventional materials. These are excellent solution for thermal barrier coatings used on turbine blades to wear resistant rotating parts and contribute a multi billion dollar economic viability.

Lubricants: The viscosity of confined nanoscale lubricant has been suggested to be the best than the same bulk lubricant. These nanotechnologically improved lubricants are recommended for microelectromechanical systems. It is also found that nanoconfined lubricants are non-Newtonian *i.e.,* shear-thinning over ranges of strain rate ($10^2 - 10^5\ s^{-1}$) for which the corresponding bulk lubricants are Newtonian. However, practical investigations on this are still in progress.

Biological: Besides the established fact that biological entities could be grouped as nanostructures there are inorganic nanoparticles for biological use. Mimics of biological weapons can be decontaminated by using ultra fine particles. Ex. Bacillus cereus spores or E. coli bacteria can similarly be disinfected adopting nanoparticle formulations. Further nanoparticles airborne and formulated from magnesium oxide and other reactive components can destroy the heat resistant Bacillus globigii spores which are mimic of Anthrase. This is expected to be a vast promising area of research and development not only for defence applications but also for medical sciences.

Imaging graphic systems: Using nanoparticulate preparations thinner imaging layers have been generated like graphic film systems. This permits higher feature performance and results in the ability to reduce the concentration of the dye component. To produce more high resolution photos, this is useful in the photographic film industry.

Electronics and computing: There are speculations that nanotechnology will create a paradigm shift in the processing architecture. Certain of such possible paradigms shifts are: quantum dot cellular automation, quantum computation, computation involving DNA strands and molecular electronics. These novel architectures will bring revolutionary changes in information technology. Miniaturization of electron devices will gradually lead to the size to that of molecule. Molecular electronics is a promising area of research. Using the batch process of chemistry molecules are designed and synthesized and then assembled into useful circuits in the process of self alignment and self organization.

Quantum computing which has resulted out of marriage between computer science and quantum physics—is another development with regard to the problem of continual miniaturization of semiconductor electronics. Although it is simple in principle, it is difficult in practical realization. It is based on the energy states of atoms and electrons, however, the elements that carry information are atoms. The idea is to derive advantage from quantum mechanics rather than be limited by it, to develop processors that simulate physical phenomenon more naturally and exponentially foster than the digital computer.

All these are developments towards the formation of a larger domain *viz.* nanoelectronics. A member of solid-state nanoelectronic switching devices are fabricated and prototype processors are already operational. A taxonomic classification of nanoelectronic devices is provided in the following table.

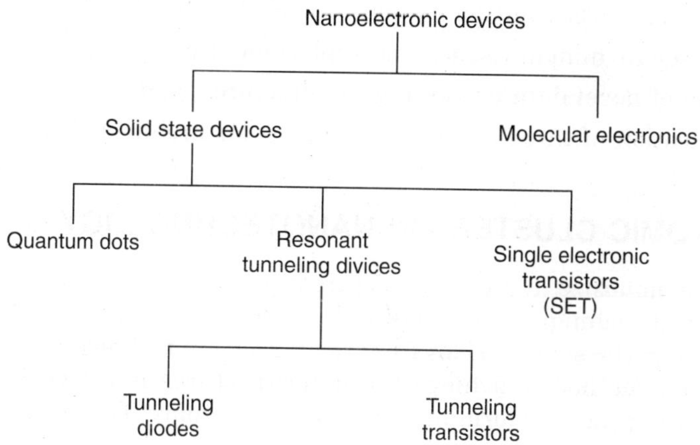

Writing with atoms: We have acquired the ability to write literally with atoms, moving the letters and placing at desired locations to form the atomic scale letters. The tininess is inexpressible ! Because the normal sized letters with this reference are about a millimeter across *i.e.*, one million nanometer. This reveals that writing at nanometer scale can accommodate one million letters into a millimeter, with this it is quite evident that how much informmation could be stored in a media of one inch dimension, which is capable of sustaining such writing.

Nanotubes: Carbon nanotubes are nature's gift to human beings in inorganic form; this is the most wonderful nanostructure that the human being has discovered. Nanotube has been established as a subject of its own. Because of the most remarkable properties like, field emission, quantum conductance, ability to form junctions and joint structures, Aharonov-Bohm oscillations, strength, stiffness, chemical robustness, and Luttinger-liquid type behaviour, etc. are some promising ones.

Next exciting application comes forth of its field emission properties. It could be possible that both single-wall and multi-wall nanotubes are used as the ultimate field emitters in the future. It has implications in obtaining high resolution displays.

Nanotubes mechanical strength enables it to be used as the tip of scanning probe microscopes; because nanotubes are far more resistant to any tip crash unlike the silicon cantilevers, and metal wires presently in use. This advantageous phenomenon is due to nanotubes slender shape and well defined tip.

Multi-wall nanotubes have been found to behave as quantum dots and most interesting results are obtained on its thermal properties. It is also found that the thermal conductivity of a multi-walled tube is more at room temperature.

Several examples can also be sited in support of the nanotechnology including medical and health, aeronautics and space exploration to defence and also national security reasons. On the whole nanotechnology provides a vision of bringing in paradigm changes in technology, the different ways materials, devices and systems are understood and also created. Certain land marks in the vision in nanotechnology can be expected as:

(*a*) Emergence of new novel phenomenon in physics and chemistry,

(*b*) Achieving orders of magnitude enhancement in computer efficiency,

(*c*) Ability to produce atomic scale documentation and capacity of storage,

(*d*) Realization of power solution for continuous presence in space,

(*e*) Ability to repair human tissues and replacement of tissue,

(*f*) Utilization of novel data processing architectures, and

(*g*) Nanofabrication of three dimensional molecular architectures.

ROLE OF ATOMIC CLUSTERS IN NANOTECHNOLOGY

Achieving the nanosize in a way is to start from the micron length scale and reduction in size using different techniques, like–ball milling and chemical methods etc. Both of these methods of reducing to the smaller sizes from the larger are technically termed as "top down" process. The another method to achieve the nanostructures is to synthesize them directly from atoms, associating them or forming clusters out of atoms. This process is called "bottom up" process.

Conclusion

Interesting intrinsic properties and their promiscuous assurance to obtain novel nanostructured materials, the field of atomic clusters has become very special and important part in the drive to advance nanotechnology. They are equally promiscuous for the studies of new fundamental properties of matter as well for creating new materials. Several advanced studies on clusters are enabling us to understand different methods of successful fabrication of nanoscale devices.

1

Nanotechnology

The prefix 'nano' in the word 'nanotechnology' means a billionth (1×10^{-9}).

Nanotechnology deals with various structures of matter having dimensions of the order of a billionth of a meter.

To get a feeling of the nanoscale

- A human hair measures 50,000 nanometeres dia.
- Bacterial cell measures few hundred nanometeres dia.
- Smallest things visible with the unaided human eye are about 10,000 nanometers in size.
- 10 H atoms in a line make up one nanometer.
- Smallest feature commonly etched on a commercial microchip today is about 130 nanometers across.

Nanoscience—the study of the fundamental principles of molecules and structures of a size between 1 and 100 nanometers.

These structures are known, perhaps uncreatively nanostructures.

Nanotechnology is the application of these nanostructures into useful nanoscale devices.

Anything smaller than a nanometer in size is just a loose atom or small molecule floating in space as a little dilute speck of vapour.

Nanostructures are not just smaller than anything we have made before, they are the smallest solid things, it is possible to make.

Though the word 'nanotechnology' is relatively new, existence of structures of nanometer dimensions is not new—existed on earth from a very long time.

Ex: Abalone, a mollusk, constructs very strong shells by organizing calcium carbonate into strong nanostructured bricks hold together by a glue made of a carbohydrate protein mix.

Cracks—initiated on the outside are unable to move through the shell because of the nanostructured bricks.

It is known that in 4th Century A.D. Roman glass makers were fabricating glasses containing nanosized metals.

An artifact from this period called the Lycurgus Cup is made from soda lime glass containing silver and gold Nanoparticles. The colour of the cup changes from green to a deep red, when a light source is placed inside it.

The great varieties of beautiful colours of the windows of mediaval cathedrals are due to the presence of metal nanoparticles in the glass.

In 1857 Michael Faraday published a paper in the 'Philosophical Transactions of Royal Society', which attempted to explain how metal particles affect the colour of church windows.

Gustar Mic—First to provide an explanation of the dependence of the colour of the glasses on metal size and kind.

Photography depends on production of silver nanoparticles sensitive to light.

Photographic film is an emulsion, which contains thin layer of gelatin containing silver halides *viz.* silver bromide—the light decomposes silver halides—producing nanoparticles of silver, which are the pixels of the image.

Richard Feynman—Nobel prize in physics—1965—contributions to quantum electro-dynamics.

In 1960—presented a visionary and prophetic lecture at a meeting of the American Physical Society, entitled "There is plenty of room at the bottom", where he speculated on the possibility and potential of nanosized materials.

He envisioned etching lines a few atoms wide with beams of electrons predicting electron beam lithography—which is used today to make silicon chips.

He produced manipulating individual atoms to make new small structures having very different properties—Indeed now has been accomplished using a scanning tunneling microscope.

He envisioned building circuits on the scale of nanometers that can be used in powerful computers.

BREAKTHROUGH IN NANOTECHNOLOGY

(1) Solar Energy

Can make solar cells efficient, cheap as newspaper, tough as asphalt road—resulting in efficient, inexpensive storage cells—low-cost power.

(2) Medicine that Cures

Patient comes to doctor—doctor asks her/him to inhale an aerosol cough, spit in a cup—sample poured to cell analyzer—in 5 min patient diagnosed—malignant cancer of the throat —in 20 minutes some 16,314 molecular markers on cancer cell are analysed.

Doctor comes out with immune machines to go after cancer cells—patient cured of malignant cells roughly in 3 weeks.

(3) Pocket Supercomputers

May not be a dream machine, with development of nanotransistors. Virtual reality systems—have begun to appear, enabling a user to walk around the image of a molecule and "touch" it, using computer—controlled gloves and goggles. Consequently computers—shrink in size and cost.

(4) Transportation Outward

Molecular manufacturing world make aerospace structures from nearly flowless, superstrong, materials at low cost—inexpensive fuel-exhaust of pure water vapour. Space flight has become clean, safe and routine.

BOTTOM-UP VERSUS TOP-DOWN

Imagine you start with a cube of gold 3 foot on each side.

Let's say we cut it into 8 little cubes, each 18 inches on a side.

The properties of each smaller cubes will be exactly the same as the properties of the big one—each little cube will still be gold, yellow, shiny and heavy—each of them will till be a soft, electrically conductive metal with the same melting point.

Again let us say—you take one 18 inch cubes and cutting into 8 pieces of 9 inches each side—still it will have same properties as of the parent cube.

Continue to cut the gold in this way—proceeding in size from feet to inches—inches to centimeters—centimeters to millimeters—millimeters to microns—still there is no change in the properties of the gold.

Eventually, we will not be able to see them with the naked eye then we need some fancy tools to keep cutting.

Still, in all the gold bricks physical and chemical properties will be unchanged.

When we reach the nanoscale—everything will change, including the gold's colour, melting point and chemical properties. The reason for this change—has to do with the interactions among the atoms that make up the gold. Interactions that are averaged out of existence in the bulk material.

Nano Gold Doesn't Act Like Bulk Gold

The last few steps of the cutting required to get the gold cube down to represent a nanoscale is referred to as "Nanofabrication" on "Nanoscale manufacturing".

- Starting with a suitcase-sized chunk of gold, successive cutting has brought it down to the nanoscale—This type of nanofabrication referred to "top-down nano-fabrication"—because we started with a large structure and proceeded to make it smaller.
- Conversely, starting with individual atoms and building up to a nanostructure is called "bottom-up nanofabrication".
- The tiny gold nanostructures—that is prepared are referred to as 'Quantum dots' or 'Nanodots'—as they are roughly dot-shaped and have diameters at the nanoscale.
- Nanoscale gold particles—can be orange, purple, red or greenish depending on their size. These red nanodots can rapidly start looking, yellow if nanodots are pushed once again together and allowing them to join.

  If nanodots are close to each other but not close enough to combine—one can see red color with the naked eye.
- In bottom-up approach—idea is to collect, consolidate and fashion individual atoms and molecules into the structure. This is carried out by a sequence of chemical reactions controlled by catalysts.

  It is a process that is widespread in biology where, for example, catalysts called 'enzymes' assemble amino acids to construct proteins—living tissue that forms and supports the organs of the body.
- In top-down approach—one starts with large-scale object or pattern and gradually reduces its dimension/dimensions.

This is accomplished by a technique called "Lithography", which shines radiation through a template on to a surface coated with a radiation sensitive resist; the resist is then removed and the surface is chemically treated to produce the nanostructures.

NANOSCALE LITHOGRAPHY (NANOLITHOGRAPHY)

- Word "Lithography" originally refers to making objects from stones.
- A lithograph is an image (usually on paper) that is produced by carving a pattern on the stone—inking the stone—then pushing the inked stone onto the paper.
- Common methods used to make current computer chips normally use optical or X-ray lithography.
- However, for nanoscale lithography—visible light cannot be employed as wavelength is at the min 400 nm.

BASIC TECHNIQUE FOR NANOLITHOGRAPHY—TOP DOWN

- Employes shining radiation through a template onto a surface coated with a radiation—sensitive resist—resist then removed.
- Surface chemically treated to produce nanostructure.

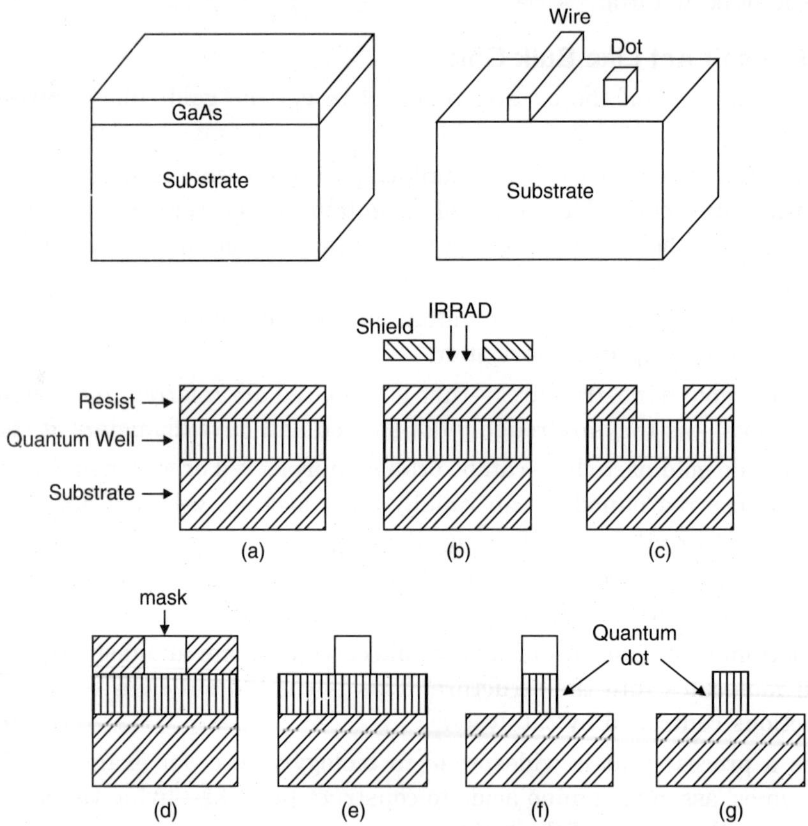

Fig.1.1 Steps in the formation of a quantum wire or quantum dot by electron-beam lithography; (*a*) initial quantum well on a substrate, and covered by a resist; (*b*) radiation with sample shielded by template; (*c*) configuration after dissolving irradiated portion of resist by developer; (*d*) disposition after addition of etching mask; (*e*) arrangement after removal of remainder of resist; (*f*) configuration after etching away the unwanted quantum-well material; (*g*) final nanostructure on substrate after removal of etching mask. [See also T.P. Sidiki and C.M.S. Torres, in Nalwa (2000). Vol. 3. Chapter 5. p. 250.]

DIP PEN NANOLITHOGRAPHY (DPN)

- Developed by Chad Mirkin and his collaboration at Northwestern University.
- DPN is similar to writing with ink using a fountain pen.
- To write lines at nanoscale—have to use nanopen.
- AFM (Atomic force microscopy) tips are ideal nanopens.
- DPN named after old-fashioned—dip pen used in school rooms in 19th Century.
- In DPN, a reservoir of "ink" (atoms or molecules) is stored on the top of the scanning probe tip, which is manipulated across the surface, leaving lines and patterns behind.

Advantages

- Almost anything can be used as nanoink.
- Almost any surface can be written on.
- One can use DPN to make almost any structure since AFM tips are relatively easy to manipulate.

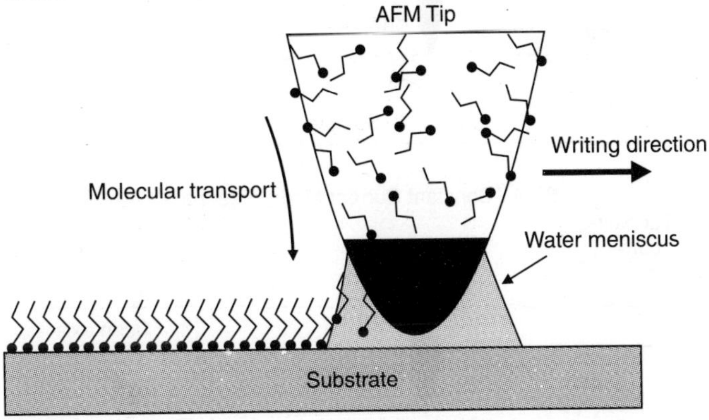

Fig.1.2 Schematic of the dip pen lithography process—the wiggly lines are molecular "ink." *Courtesy of the Mirkin Group, Northwestern University*

Steps: To place a radiation—sensitive resist on the surface of the sample.

- Sample is then irradiated by an e^- beam in the region where the nanostructure will be located. Radiation chemically modifies the exposed area of the resist so that it becomes soluble in a developer.
- Application of the developer to remove irradiated portions of the resist.
- Insertion of an etching mask into the hole in the resist.
- Lifting off the remaining parts of the resist.
- Areas of quantum well not covered by the etching mask are chemically etched away to produce the quantum structure.
- Finally the etching mask is removed—to provide desired quantum structure—which might may be the quantum wire/quantum dot.

E-BEAM LITHOGRAPHY

- Light based industrial lithography is limited to creating structures no smaller than the wavelength used.
- Smaller wavelength light has higher energy, so it can have unwanted side effects like it can blow the structure, which you are trying to create similar to something watering garden plants with a fine hose.

- Alternate is to use electrons instead of light
- E-beam lithography—used to make structures at the nanoscale.
- Has applications in current microelectronics manufacturing and is one approach to make them.

Achieve this, the probe tip moves up and down resulting in a pattern and is recorded.

Constant-Probe Height

- The height of the probe tip is maintained constant, the resulting variations in current is recorded and surface pattern is produced.

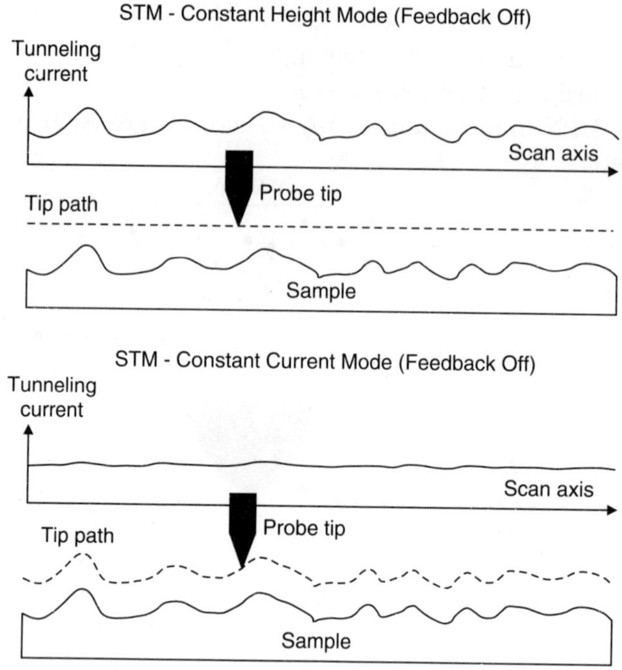

Fig. 1.3 Constant-height (top sketch) and constant-current (bottom sketch) imaging mode of a scanning tunneling microscope. [From T. Bayburt, J. Carison, B. Godfrey, M. Shank-Retzlaff, and S. G. Sligar, in Nalwa (2000), Vol. 5, Chapter 12, p. 641.]

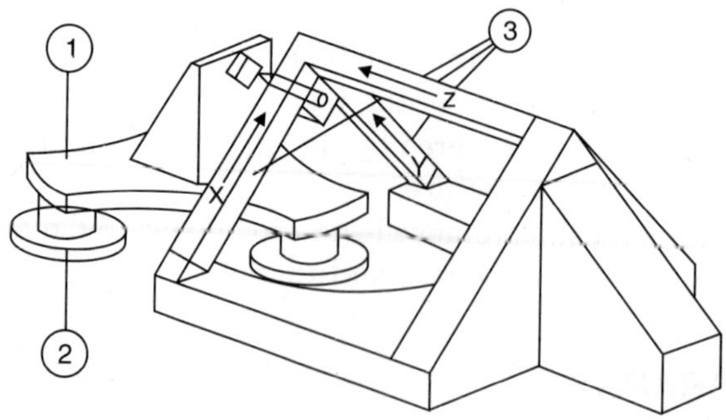

Fig. 1.4. Scanning mechanism for scanning tunneling microscope, showing (1) the piezoelectric baseplate, (2) the three feet of the baseplate and (3) the piezoelectric tripod scanner holding the probe tip that points toward the sample. (From R. Wiesendanger, *Scanning Probe Microscopy and Spectroscopy,* Cambridge Univ. Press, Cambridge, UK, 1994, p. 81.)

The STM often employs—piezoelectric tripod scanner. A piezoelectric is a material in which an applied voltage elicits a mechanical response and produces electric current in response to mechanical pressure.

Applied voltage induce piezo transducers to move the probe.

TOOLS FOR MEASURING NANOSTRUCTURES

- Tools which helped in nanoscience revolution firstly, were the so-called scanning probe instruments.
- 'Scanning probe instruments' based on the idea first developed at the IBM Laboratory in Zurich in the 1980s.
- Idea is simple—If one rubs finger along a surface, it is easy to distinguish velvet from steel or wood from tar. Different materials exert different forces on the finger as it is dragged on the surface. Here finger acts like force measurement structure.
- In scanning probe measurements—probe—also called tip, slides along a surface in the same way as finger does. The probe is of nanoscale dimensions, often only a single atom in size where it scans the target.

As probe slides, it can measure several different properties, each of which corresponds to a different scanning probe measurement.

SCANNING TUNNELING MICROSCOPE (STM)

- A STM utilizes a wire with a very fine point.

This fine point is positively charged and acts as a probe when it is lowered to a distance of about 1 nm above the surface under study.

Electrons at individual surface atoms are attracted to the +ve charge of the probe wire and jump/turned up to it—thereby generating a weak electric current.

The probe wire is scanned back and forth across the surface in a master pattern, in either a constant-height mode, or a constant-current mode.

Constant-current Mode:

- A feedback loop maintains a constant electric current—to scanning probe in nanometer increments along x, y and z directions.
- The tunneling current, which varies exponentially with the probe—surface atom separation, depends on the nature of the probe tip and the composition of the sample surface.

ATOMIC FORCE MICROSCOPE (AFM)

Fundamental difference between STM and AFM:

- Former monitors electric tunneling current between the surface and the probe tip, where as the latter monitors the force exerted between the surface and the probe tip.

- AFM, similar to STM—has two modes of operation
- AFM can operate in a close contact mode—core-to-core repulsive forces with the surface dominate.

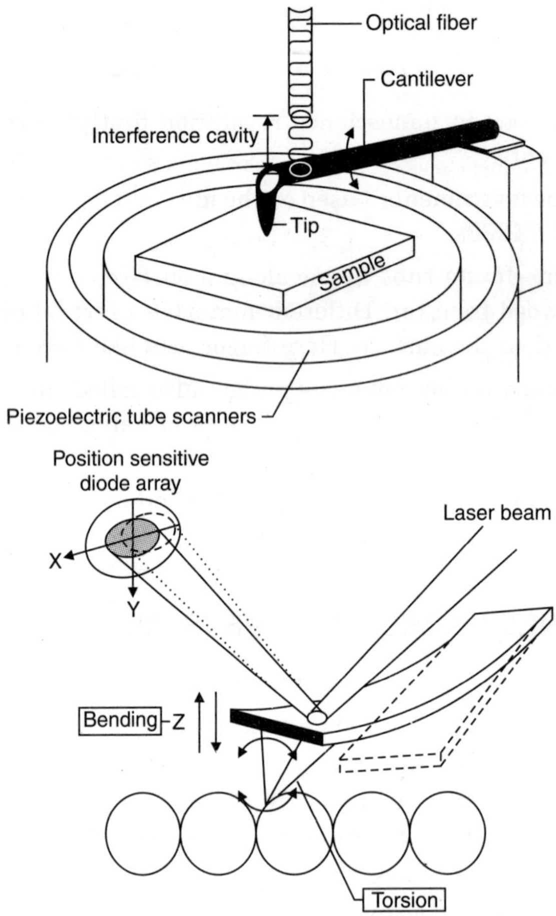

Fig. 1.5 Sketch of an atomic force microscope (AFM) showing the cantilever arm provided with a probe tip that traverses the sample surface through the action of the piezoelectric scanner. The upper figure shows an interference deflection sensor, and the lower enlarged view of the cantilever and tip is provided with a laser beam deflection sensor. The sensors monitor the probe tip elevations upward from the surface during the scan

- Non contact mode—relevant force is the gradient of the vander Waals potential.
- Like in STM case—piezoelectric scanner is used.
- Vertical motions of the tip during the scanning may be monitored by

  (*a*) interference pattern of a light beam from an optical fiber.

  (*b*) the reflection of a laser beam.

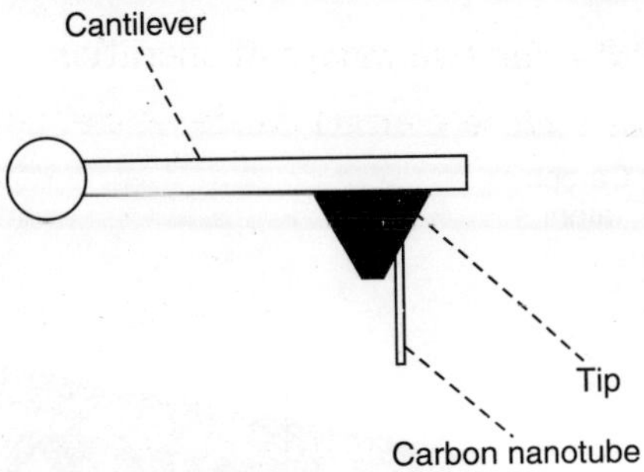

Fig. 1.6 Illustration of a single-walled carbon nanotube mounted on an STM tip attached to a cantilever arm of an atomic force microscope. [Adapted from H. Dai et. al., *Nature* 384, 147 (1996).]

- AFM employs a sharp tip mounted on a cantilever spring, which is scanned closely over the surface of a material.
- The deflection of the cantilever is measured.
- In the region of surface atoms the deflection is larger because of the larger interaction between the tip and the atoms.
- Cantilevers are fabricated by photolithographic methods from silicon, silicon oxide or silicon nitride.
- They are typically 100 μm long and 1 μm thick and have spring constants between 0.1 and 1.0 N/m (Newton per meter)
- Tip material is usually made of carbon nanotube.
- A multiwalled nanotube (MWNT) is bonded to the side of a tip of a conventional silicon cantilever using a soft acrylic adhesive.
- If nanotube crashes to the surface—generating a force greater than the Euler buckling force, the nanotube does not break, but rather bends away and then snaps back to its original position.
- Such MWNT tip can also be used in the tapping mode.

NANOTUBES

What is Nanotube: Almost always carbon nanotubes, referring to the wires of pure carbon that look like rolled sheets of graphite or like carbon soda straws.

Carbon nanotubes were discovered in 1991 by Sumio Iijima, a researcher at NECs. Fundamental Research Lab.

Special Characters of Carbon Nanotubes

- One of the most commonly mentioned building blocks of nanotechnology.
- Has one hundred turns the tensile strength of steel.
- Thermal conductivity better than all but the purest diamond.
- Electrical conductivity similar to copper, but with ability to carry much higher currents.

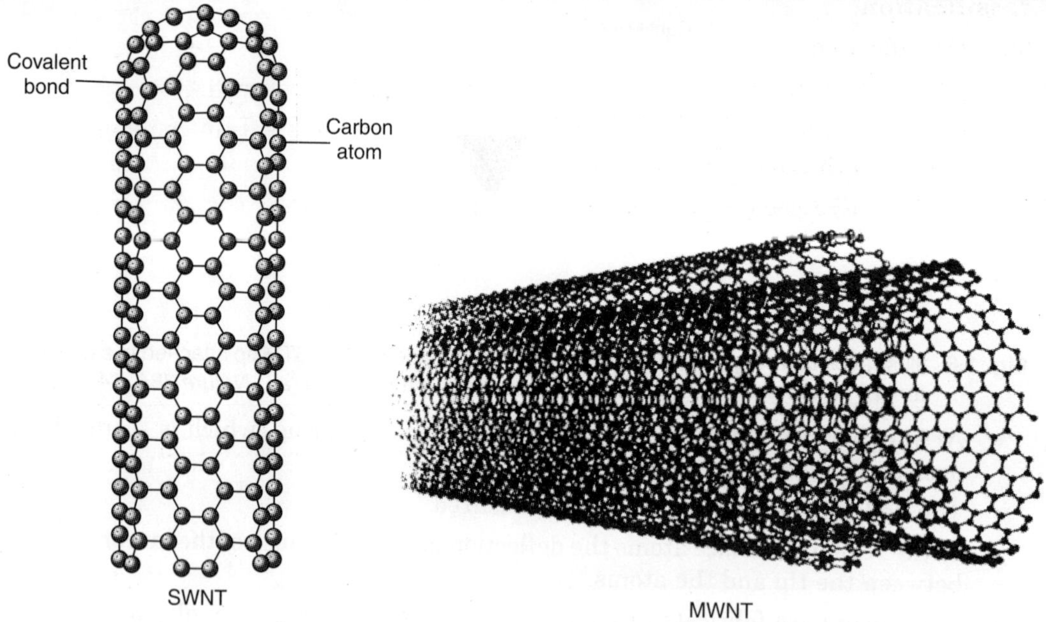

Fig. 1.7 Illustration of a multiple-walled carbon nanotube

- Though stronger to steel, density is about one quarter that of steel.
- Nanotubes are strong but are also elastic. This means it takes a lot of force to bend a nanotube, but it will spring back to its original shape when you release it, just like rubber band.
- Carbon nanotubes are a little bit sticky. The electron clouds on surface of each nanotube provide a mild attraction force between the nanotubes. A carbon tube has character of a nonpolar molecule.

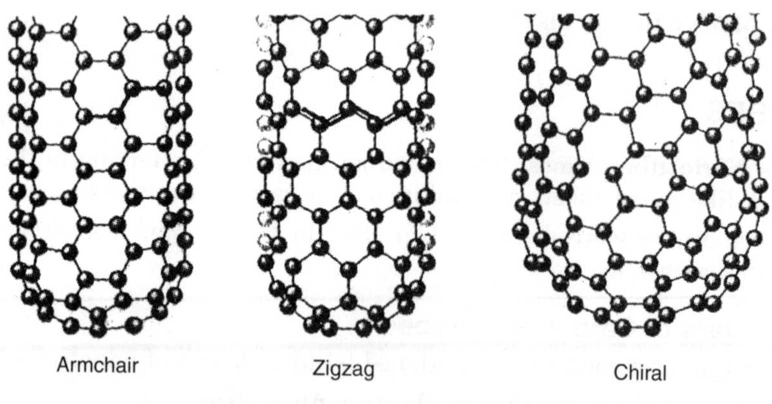

Fig. 1.8 Armchair, zigzag, and chiral nanotubes

Nanotubes come in a variety of flavours: long, short, single-walled, multi-walled, open, closed, with different types of spiral structure. Each type has specific production costs and applications.

Some have been produced in large quantities, others are only now being produced commercially with decent purity and in quantities greater than a few grams.

Classification

Major classification

- Single-Walled Carbon Nanotubes—SWNTs—have cylindrical wall.
- Multi-Walled Carbon Nanotubes—MWNTs—have cylinders within cylinders.
- Lengths of both types vary greatly.
- Generally microscopic rather than nanoscopic.

SWNTs

- Stars of nanotube world.
- Much harder to make than the multi-walled variety.
- Basically tubes of graphite and are normally capped at the ends.
- Theoretical minimum diameter of a C-nanotube is around 0.4 nm.
- More pliable than MWNTs, can be twisted, flattened and bent into small circles or around sharp bends without breaking.
- Volumes of SWNTs produced are currently small and the quality and purity are variable.

MWNTs

- Basically like Russian dolls made out of SWNTs—concentric cylindrical graphite tubes.
- Typically 100 times longer than they are wide and have outer diameter mostly in the tins of nanometers.
- Though easior to produce significant quantities of MWNTs, compared to SWNTs, their structures less well understood than SWNTs.
- Found in different shapes and arrangements like bamboo trunks, sea urchins, recklaces/coils etc.
- Most of nanotube applications—considered based on MWNTs—because they are easier to produce and at reasonable price.

Producing Nanotubes

Three methods are tried by various companies:

(I) High pressure carbon monoxide deposition /HiPCO: Involves heated chamber—through which CO gas and small clusters of iron atoms.

When CO lands on iron clusters—iron acts as a catalyst and helps CO molecule to break into carbon atom and an O_2 atom.

C atom bonds with other carbon atoms to start nanotube latice.

O_2 atom combines with another CO molecule to form CO_2 which then floats off into the air.

(II) Chemical vapour-deposition/CVD: A hydrocarbon—methane gas—flows into heated chamber containing a substrate coated with a catalyst like iron particles.

Temperature in chamber high enough to break bonds between C atoms and H atoms of CH_4—resulting in C atoms attaching to catalyst particles—where they bond to other C atoms—forming nanotube.

(III) PLASMA Process: CH_4 used as source of carbon—passed through a plasma torch—but no greater details have been revealed by any industry. Said to be 25 times more efficient compared to other two methods.

Applications

- Bucky tubes have extraordinary electrical conductivity, heat conductivity and mechanical properties.
- They are probably, best electron field-emitter.
- They are polymers of pure carbon and have fine nanotube structure, which give extraordinary characteristics to nanotubes with tremendous applications.

Field emission: Best known field emitters, because of their high electrical conductivity and sharpness of their tip.

Used in cold-cathode lighting sources, lighting arrestors and electron microscope sources.

Conductive plastics:

- Plastics are being famously known for their good electrical insulation.
- Now plastics are loaded c⁻ conductive fillers the nanotubes.
- Used for shielding composites, gaskets, antistatic materials and radar absorbing materials.

Energy storage:

- Bucky tubes have tremendously high surface area, good electrical conductivity, very important linear geometry.
- Used in lithium-ion batteries, supercapacitor electrodes, in variety of fuel cell components.

Fibres and fabrics: Fibres spun of pure buckytubes are being developed. Such super strong fibres will have applications including body and vehicle armour, transmission line cables, waven fabrics and textiles.

Sensors: To detect chemical vapours. Molecular of chemical vapour would land on the nanotube and forms covalent bonds with C atoms, thus changing the electrical conductivity of nanotude either by increasing or decreasing electrical conduction which can be sensed.

Fuel cell-powered motors:

- One can look at possibility of storing hydrogen in nanotubes, can be used as a fuel tank for H-fuel cell-powered cars.
- Researchers are looking to use nanotubes to make transistors, memory cells and wires.
- Recently nanotubes are being explored in biomedical applications. Cells have been shown to grow on bucky tubes without nanotubes having any toxic effect on them.
- Also used to prepare vascular slints, for neuron growth and regeneration.

■ BUCKY BALLS

- Bucky ball short form for Buckminster fullerene.
- Discovered through an interesting collaboration of researchers from two universities—they were Curl, Smalley and Kroto—1985.
- Later produced in large amounts by Krät schmer and Huffman—1990.
- Is a molecule containing 60 carbon atoms. Each carbon atom is bonded to 3 adjacent carbon atoms like in graphite.

- Bucky ball form a sphere that's about 1 nm in diameter.
- Actual the smallest fullerene is the C_{60}.
- Also found in the natural environment—only in the ppm range.
- Some places Shunga in Russia, New Zealand. Sudbury in Canada.

Chemical and physical properties:

- Carbon atoms in fullerens are sp^m. The free valence electrons are in the π electron system. π electron system influences chemical reactions of the fullerens. They show aliphatic behaviour.
- Unsoluble in water and stable in air.
- Solvents used are CS_2, *o*-dichlorobenzene, toluene and Xylene.
- This layers of fullerens—coloured—yellow to yellow green.
- Heating without air up to 1500°C—transform to graphite.

Creating Bucky Balls

- Smalley created bucky ball by vapourizing carbon with a laser and allowing C-atoms to condense.
- This technique produces small amounts—just show to existence but cannot understand capacities.
- Later researchers at Max Planck Institute and University of Arizona vapourized carbon by placing two carbon electrodes close together and generating an electric arc between them in a reaction chamber filled with a low pressure of helium or neon.
- Method produced large quantities of bucky balls.

Uses:

(1) **As antioxidants**

- Neutralize free radicals, when bucky ball meets a free radical, the unpaired e^- in the free radical pairs up c^- bucky balls delocalized e^-s, forming a covalent bond between free radical and C atom.

(2) **In medical imaging and drug delivery**

- Metal gadolinium—used for medical imaging—but some amount may retains in body and can cause damage.
- Trying to contain this in bucky balls—researchers hope to flush the metal—enclosed in a bucky ball combination completely from the body.
- Use of bucky balls to deliver drugs directly to infected regions of the body. Usually affected regions differ in pH from the healthy ones.
- Researchers hope to design bucky balls—to release drug—sensing change in pH.

(3) **Polymers**

Dupont and Exxon are trying to develop stronger polymers.

(4) **Antioxidant type drugs**

- Antiaging or anti-wrinkle creams are being developed by Mitsubishi.
- Taiwan University in testing bucky ball based drug to fight asteriosclerosis.
- C sixty working on both burn creams and an HIV drug.
- Sony—trying to develop efficient fuel cell membranes.
- Siemens has developed a bucky ball—based light detector.
- Seagate—using bucky balls to develop diamond—hard coatings for computer disk driver.

STRUCTURE—PROPERTY RELATIONSHIPS AT THE NANOSCALE

Polymer Cross-linking

- Polymers include all the plastics, with which we are familiar.
- Composed of monomers and are quite large. Monomers repeat in variety of ways to build polymers of various sizes.
- Have a wide range of properties—dependent on their molecular structure.
- One of the important structural quality, which affects behaviour is cross-linking.
- Polymers composed of cross-links—more rigid than noncross-linked polymers.
- Higher the number of cross-links greater is the capacity to withstand physical stress and are more rigid.

Poisson's Ratios

- When most objects are stretched, they become thinner in cross-section.
  Ex. Rubber band, chewing gum etc.
- Represented by Poisson's Ratio *i.e.,* μ(mu)

$$\mu = \frac{\%\text{ in width}}{\%\text{ increase in length}}$$

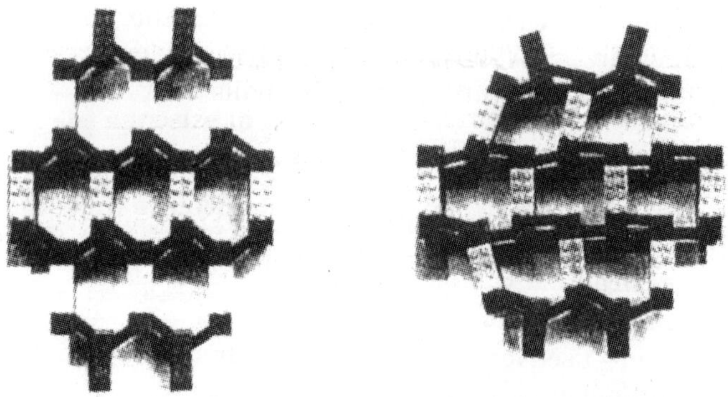

Fig. 1.9 A LEGO® lattice with a positive Poisson's ratio: (left) unstretched and (right) stretched horizontally. Note that the lattice contracts vertically (in a direction perpendicular to stretching)

Fig. 1.10 A LEGO® lattice with a negative Poisson's ratio: (left) unstretched and (right) stretched horizontally. Note that the lattice expands vertically (in a direction perpendicular to stretching)

Silicates

- Silicates and aluminisilicates, large classes of minerals found in many forms throughout the earth's crust.
- Variety of minerals based on building units of silicate ions *i.e.,* SiO_4^{4-} (Aluminium silicates AlO_4^{5-}) SiO_4^{4-} —shaped tetrahedra c^- Si at the center and oxygen atoms at the four corners.
- These tetrahedral units—connect at their corners of form chains, rings, sheets and 3-dimensional structures.

Asbestos type minerals—consist of silicate chains arranged parallel to each other. Interactions are weak, resulting in fibrous structure. But show high heat resistance, hence asbestos used as insulating materials.

Mica and Clay type minerals—consist of aluminosilicate sheets arranged perpendicular to each other.

Interactions are relatively weak, allowing mical minerals to bc peeled/split into flat sheets.

Vermiculite—mica type minerals—is "popped" like popcorn due to evaporation of water trapped between its sheets, resulting in a fluffy structure used as packing material and insulation.

- In some cases, weak interactions between aluminosilicate sheets allow sheets to slide past each other. Therefore, resulting in slippery feel of talc and clay (wet).

Zeolites—Complex, 3-dimensional aluminosilicate structure, nanoporous solids. Also contain channels and empty spaces throughout their volume resulting in a porous, high surface area structure.

They can trap specific molecules, and therefore used as ion exchanges, molecular sieves and catalysts.

Metal deformation:

- Metals deform under applied pressure
- Sheets of atoms slide past each other
- Interface along c^- sheets slide are called slip planes.
- Metal sheets that are flat, move past each other more easily than sheets which are wrinkled/distorted. That means, metals with many defects in their crystal structure are less easily deformed than metals with relatively few defects.

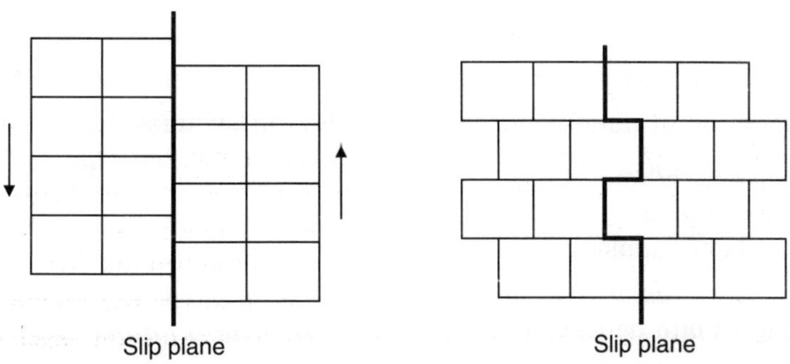

Fig. 1.11

Composites

- Is a substance composed of two/more materials with different base structures, combined in such a way that the end product has different properties than either of the parent materials.
- They are two distinct structures blended together, resulting material has better properties than either parent material.
- Reinforced concrete material, found in bridges, buildings is an example.
- Composite bodies reduce rusting effects and composite dashboards improve safety. Composites are being developed using nanoscale components.

Giant Magnetoresistance (GMR)

- Is a property in which a magnetic field affects how well a material conducts electricity.
- 'Read' heads in computer hard drives are made out of GMR materials due to the increasing magnetic sensitivity of these heads over old aged heads.
- One type of GMR—called a magnetic multilayer material—composed of alternating layers of ferromagnetic material and nonmagnetic metals, each having thicknesses on the nanoscale.
- Here the resistance of certain materials drops dramatically as a magnetic field is applied.
- It is describe as Giant since it is a much larger effect than had ever been previously seen in metals.
- This has lead to great interest shown by physicists, and eventually huge technological applications were tapped as 'magnetic recording and sensors'.
- IBM were first to market with hard disks based on GMR technology.

NANOBIOTECHNOLOGY

- Is the application of nanotechnology to the life sciences.
- It is the intersection of inorganic and organic engineering to solve critical problems in biology. These problems can be the creation of
  - → New drugs,
  - → Drug delivery vehicles
  - → Diagnostics
  - → Sensors
  - → Assays
  - → tools such as fluidics,
  - → and manufacturing process for all the above.
- This field includes two approaches
  - (*a*) Application of nanoscaled tools to biological systems
  - (*b*) Use of biological systems as templates in the development of novel nanoscaled products.
- Why Nanobiotechnology ?

  Mother nature did it first and still continues to do it. It has been doing things of these kind (if one believes of evolution) from several million years.

→ Nature applies nanotechnology daily to grow the multifunctional cells and tissues of plants and animals from a single biological cell.

→ A cell is a warehouse of nanoscale machines.

→ Biology can teach the physical world of electronics, computing material sciences and manufacturing. Ex: how fish swims, Honey bee buliding strong hive, Insects flying.

What is important? Is there exist biomolecular analogues of conventional functional devices?

Ex:		
	(*a*) Vehicles	(*a*) Haemoglobin
	(*b*) Assembly	(*b*) Ribosomes
	(*c*) Motors, Generators	(*c*) ATP Synthases
	(*d*) Train tracks	(*d*) Action filament networks
	(*e*) Train controlling centre	(*e*) Centrosome
	(*f*) Digital databases	(*f*) Nucleosomes
	(*g*) Copy machines	(*g*) Polymerases
	(*h*) Chain couplers	(*h*) Ligases
	(*i*) Bulldozer, Destroyers	(*i*) Proteases, Protiosomes
	(*j*) Mail sorting machines	(*j*) Protein sorting machines
	(*k*) Electric fences	(*k*) Membranes
	(*l*) Gates, keys, passes	(*l*) Ion channels
	(*m*) Internet nodes	(*m*) Neuron synapses.

Nanobiotechnolgy Applications

(1) **Nanomedicine**

→ Nanoparticles would deliver treatments to specifically targeted sites c^- help like nanoshells.

→ Improved imaging Ex: Quantum dots.

→ Nanoscale modifications of implant surfaces.

(2) **Nanobiosensor (Biochip)**

→ Translating Biomolecular recognition into nanomechanics.

Biological samples can be screened for the presence of specific DNA. Cantilevers of AFM empolyed. Surface cantilever is coated with DNA which enables to bind to a particular DNA sequence.

→ Bioelectronic sensor

Provided with thin surfactant polymeric electrically conducting layer. On this, bound are specific binding pairs. Binding of an analytic or a reagent to this specific binding pairs change the electrical, optical or structural properties of the layer for measurement of reagent. This is a very sensitive instrument.

(3) **Molecular Self Assembly (MSA)**

→ a powerful method for assembling atomically precise materials and devices.

Biological organisms are composed of molecular building blocks *viz.* nucleic acids, phospholipids and are equipped to assemble into well organized structure.

→ This is being exploited to build nanoscale machinery and microelectronics.

Ex: Smart plastic—that assemble themselves into photonic crystals useful for data storage and telecommunications.

(4) **Intelligent Drug Delivery System:** Nano gate implants—Hast drug reservoir internal—has a nanopore membrane and a outer protective screen.

Drug gets released through nanopore membrane at a constant rate. The entire capsule is referred to as polymer capsule.

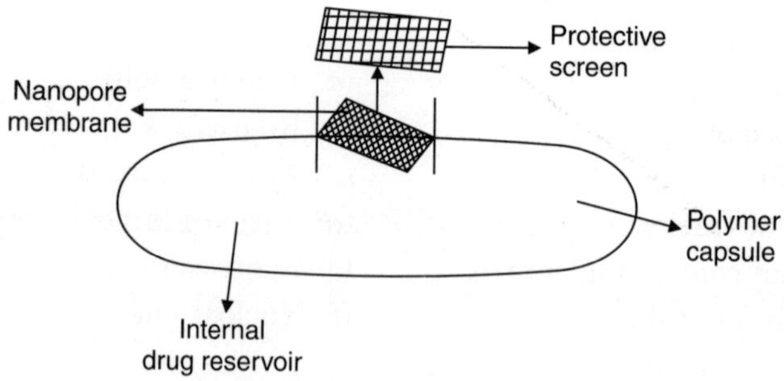

Fig. 1.12

(5) **Nanomachines/Nanorobots:** Hypodermic syringe less than 0.02 inch in diameter would inject thousands of nanorobots into blood vessel.

- Nanorobots will have small dish to send and receive data from outside body.
- Incredibly small rotor saw would scrape the plaque formed by cholestered choking the blood vessel.
- Constantly perform the task or programmed for biodegradation.

Future

Engineers in future shall have expertise both in biology and technology.

(1) Engineered biomolecular machines.

(2) Nanomedical surgical implants.

(3) Molecular healing and repair of injury and disease.

(4) Smart drugs.

(5) NEMS.

(6) Mobile nanopharmacies, Nanomachines etc.

MATERIALS CHARACTERIZATION TECHNIQUES

Several techniques and instruments are employed to study the surface architecture and other characters of the materials under study.

The following are the whole range of gamut of equipments employed in material characterization technique.

Atomic Force Microscopy—AFM

- It provides 3-dimensional topographic information about a sample by probing its surface with a very sharp tip.

- Tip is scanned laterally across the surface, and the vertical movements of the tip are recorded and used to construct a quantitative 3-dimensional topographic map.
- Lateral resolution of the image can be as small as the tip radius (5–15 nm), and the vertical resolution can be in the order of angstrom.

Energy Dispersive X-ray Spectroscopy (EDS)

- Is a standard procedure for indentifying and quantifying elemental composition of sample in small areas as small as few cubic micrometers.
- Characteristic X-rays are produced when a material is bombarded with electrons in an electron beam instrument such as scanning electron microscopy.
- Detection of these X-rays can be accomplished by an energy dispersive spectrometer, which is a solid state device that discriminates among X-ray energies.

Scanning Election Microscopy—SEM

- Uses a focused electron beam to scan small areas of solid samples.
- Secondary e^{-}s are emitted from the sample and are collected to create an area map of the secondary emissions.
- The intensity of secondary emission is very dependent on local morphology, the area map is a magnified image of the sample.
- Spatial resolution is as high as 1 nm in sophisticated instruments, but most give 4 nm resolution.
- Magnification factors can exceed 5,00,000 times.
- Typical applications include
  (*a*) Studying surface topography of solids.
  (*b*) Grain size—powders/solids.
  (*c*) Particle size and indentification.
  (*d*) Determination of fracture modes.
  (*e*) Reverse engineering.
  (*f*) Corrosion products/pitting.

Transmission Electron Microscopy—TEM

- A high voltage 120–200 keV finely focused e^{-} beam is passed through a thin (50–200 nm) solid sample.
- Contrast is derived by e^{-}s scattering from atoms in the material.
- Electrons undergo coherent scattering *i.e.*, diffraction from lattice planes yielding paterns and identification.
- Characteristic X-rays are produced and detected in a separate detect permitting qualitative elemental analysis.

Electrode Sputtering

- In an evacuated chamber, an argon partial pressure of approximetal 0.05 m bar is established through a gas dosing valve, and a glow discharge is ignited between an anode (specimen) and cathode (metal target) by means of high voltage.
- The resulting positiviely charged argon ions are accelerated toward the cathode and through their impingement, knock metal atom out (sputter) of the target.

- Numerous collisions occur between sputtered metal atoms and the gas molecules in the process chember, resulting in strong scattering of the metal atoms.
- These are deposited uniformly on the specimen(s) in the chamber.
- Even on strongly fissured specimen surfaces, a hamogeneous thin metal film is deposited. The conductivity of which is sufficient for scanning e^- microscopy specimens.
- The conductivity depends not only on the target material but also on parameters such as working distance, gas pressure, glow current and process duration.

Applications

(1) Conductive coatings for e^- microscopy.

(2) Electrode deposition for electrical testing.

Metals employed

Au, Ag, Pt.

High Temperature Furnace Lab Description

- This facilitates the development and characterization of ceramic material.
- High temperature box furnaces capable of producing 1,650°C.
- Mid range box furances capable of producing 1,400°C.
- Lower range box furances capable of producing 1,100°C.
- Usually furances controlled with centralized computer, through which one can program any temperature and record the profile achieved.

Ion Chromatography

- Used for chemical analysis.
- The chromatographic column allows for separation and a conductivity cell for detection.
- Common inorganic anions in solution are routinely determined on this instrument.
- Additional columns are employed to determine more complex in ions instrument.
- A UV/visible detector is used to determine small groups of organic species in solution such as—alcohols, carboxylic acids, amines etc.
- Samples, usually filtered before injection.
- Strong matrices *viz* sea water should be avoided.
- Typically 2–3 ml of sample solution is required.

UV/Visible Spectroscopy

- A beam of visible and/or UV light either passes through a liquid sample or is reflected off a solid sample.
- The intensity of the light as a function of wavelength is compared to a reference absorbance/transmittance spectrum.
- Different species in a sample absorb light at characteristic wavelengths.
- The degree of absorption is proportional to the molar concentration of the species present in the sample.
- Wavelength range: 190–1,100 nm; Slit Width: 1nm

X-Ray Diffraction

- X-rays c^- specific wavelength, on the order of lattice spacing are scattered from the atomic plances in a crystalline material yielding diffraction peaks.
- Resulting diffraction pattern—used to identify unknown crystalline phases, determine residual stresses, preferred orientation/grain size.
- Powder samples of 325 mesh employed.
- Thin film samples 1.125″ × 1.125″ × 0.5″
- Usually samples mounted on aluminium barrels, special arrangements are made for some other samples.

Advantages

(1) Crystalline phase identification (2) Crystalline *Vs* Amorphous (3) Non-destructive

Application

Mineral analysis in rocks, clays, refractories, pharmaceuticals, alloys, cements and industrial dusts.

■ MICROELECTRONIC FABRICATION

- Microelectronic engineering is the area of technology associated with the design and fabrication of electronic devices/systems or subsystems using extermely small components the "integrated circuits."
- MEMS (Micro-Electro-Mechanical-Systems) is the integration of mechanical elements, sensors, actuators and electronics on a common silicon substrate through 'Microfabrication technology'.
- Electronics are fabricated using 'integrated circuit' (IC) process like CMOS or BICMOS process (Bipolar complementary metal oxide semiconductor).
- On the other hand Micromechnaical components are fabricated using compatible "Micromachining" process, that selectively etch away parts of the silicon wafer or add new structural layers to form the mechincal and electromechincal devices.
- MEMS promises to revolutionize by bringing together silicon-based microelectronics with micromachining technology, making possible the realization of complete systems-on-a-chip.
- MEMS is an enabling technology allowing the development of smart products. Here microelectronic integrated circuits can be thought of as the "brains" and its decision making capability with "eyes" and "arms". This will allow microsystems to sense and control the enviroment.
- "Sensors" gather information from the environment through measuring mechnical, thermal, biological, optical and magnetic phenomena.
- The electronics then process the information derived from the sensors and through some decision-making capability direct the actuators to respond by moving, positioning, regulating, pumping and filtering, thereby controlling the environment for some desired outcome or purpose.

- Another important thing about MEMS is that these devices are manufactured using "batch fabrication techniques", something simillar to adapted in ICs, this allows for producing MEMS with high reliability and sophistication and functionality on a small siliconchip at a relatively low cost.
- A typical MFF laboratory may occupy an area of few hundred square meters (750–1,000) depending on the volume of MEMS production this requirement may go up drastically.
- These labs require class 100 environment.
- The lab/production unit will have many functional modules for the preparation of MEMS.
- Lithography: Deep UV, Extreme UV, X-ray.
- Etching: Dry etching → Material is sputtered/dissolved using reactive ions.
  Wet etching: For silicon and metals, material is dissolved when immersed in a chemical solution.
- Diffusion and Ion implantation.
- Thin film deposition—chemical vapour depositon (CVD), Plasma Enhanced CVD, Epitaxy CVD.

Other facilities

- Testing, dicing of wafers, packaging, bonding, yield theory and measurement.
- Measurement techniques—optical microscope, scanning e^- microscope, energy dispersive analysis by X-rays, Augue analysis, secondary ion mass spectroscopy (SIMS), laser ion mass spectroscopy (LIMS), Rutherford backscatter spectroscopy (RBS), X-ray diffraction etc.
- Silicon foundaries.

LIMITS OF MICROFABRICATION

- Molecular machine systems, common in biology, can become a basis for a new style of physical technology.
  → New products and system proposed include
  (*a*) Nanometer scale structures
  (*b*) Atomic precision
  (*c*) Low defect densities and
  (*d*) High manufacturing productivity.
- Physical technology means our ability to fabricate physical objects, they may be
  (*a*) Scientific instruments.
  (*b*) Consumer products.
  (*c*) Tools to fabricate other physical objects.
- To understand, ultimate limits in the use of physical technology, we must understand limits of fibriaction.
- Physical law
  → Discussions on technological possibilities can be obtained by applying models of physical law, which can give technical proposals that fall into 3 categories.

(*a*) Those that are physically feasible.
(*b*) Those that are physically infeasible.
(*c*) Those for which physical feasibility cannot be determined.

Therefore, model of physical law can be applied and an attempt can be made regarding the possible implications.

- Fabrication process
  - → To study fabrication processes, or needs to take account of the complexities of condensed matter structures and their transformations.
  - → Important point to be noted here is that, physical law models are applied only for ordinary matter that are accessible and available in this planet, and does not involve exatic matter, monopoles/pressures and densities found inside neutrons.

CONVENTIONAL FABRICATION

- Limitations in conventional fabrication are because of the inability to directly control the motions of atoms and molecules during the formation of product objects.
- Some of typical conventional processes

(*a*) **Inexact Structures**

- Many macroscopic objects, made by shaping materials either by
  (*i*) Subtractive process *viz.* machining
  (*ii*) Near net shape process *viz.* moulding
  (*iii*) Additive process *viz.* stereolithography.
- These macroscopic processes are imprecise on an atomic scale, providing only approximate control.
- Materials themselves are imprecise, some of it are granular and some are amorphous.
- Microlithographic process are similar to macroscopic processes and involve either volumetric or surface diffusion and usually because of fluxes they are distributed over microscopic area and give no precise control.

(*b*) **Exact Structures**

- Chief technology for producing complex, atomically precise structures, today is organic synthesis.
- By controlling macroscopic variables *viz.* composition and temperature during reaction and purification processes, chemists can build atomically precise structures.
- However this technology, face several basic difficulties
  (*i*) technology is relying on diffusion through solution. Therefore atleast one component must be soluble.
  (*ii*) Each substructure designed must show strong reactivity with only a single partner-otherwise reduces yield
  (*iii*) Usually, nearly 200 sequential steps are involved and converts 10^3 kg of reagents into 0.001 gm of product.

(*c*) **Limitations of Current Fabrication Processes**

- There are many short comings
- A 'diamondoid' structure that can occupy a volume of one cubic nanometer is estimated to be 10^{148}.

- The ultimate shape of this product can be chosen randomly.
- Most of these structures are irregular and asymmetrical. Therefore, cannot be made by these techniques to crystals.
- The technology, use today to make crystals is organic synthesis, but still yields very irregular structures and even do not have 100 atoms.

MOLECULAR MANUFACTURING

- Molecular manufacturing—Mechanically guided chemical synthesis—'Mechanosynthesis'.
- Principles of physics indeed permit to precisely manoeuver atoms and molecules and to put in the place, where chemists says.
- Practical molecular manufacturing processes, must use machines with components of microscopic scale, preferably built with atomic precision. These machines are themselves candidates for production using molecular manufacture.
- These machines, have substantial complexity *viz.* six-axis robotic positioning machanism.
- IBM used 35 Xenon atoms to spell 'IBM', by positioning tip of scanning tunneling microscope, however this technology could not produce stable structures of practical interest.

Physical Limits of Fabrication

- Fundamental physical limit on what can be fabricated appear to coincide with physical limits on what can stably exist.
- At the moment, condensation is a seldom fabrication technique to produce objects more complex than fullerenes.
- Moreover other constraining factor is said to be the amount of time of material.
- Certain structures have large internal stresses and can be unstable.
- Other structures of low stability *viz.* liquids, gels do not lend themselve for precise structural definition or step-wise construction therefore in these cases limits of fabrication is clear.

BIOMOLECULE SURFACE INTERACTION

Studying fundamental concepts about biomolecules and cells on surfaces is very important.

For example: In microarray technology, surfaces must be designed and prepared to optimize the immobilization of probe biomolecules and/or cells and at the same time it should resist non-specific binding of target species.

- In this discussion, we will try to understand fundamental concepts underlying DNA, protein, small biomolecule and cell attachement to surface.
- Biomolecule attachment is dependent on the properties of
  (*a*) Biomolecular surface
  (*b*) Solid surface
  (*c*) Liquid medium

of the above three property exhibited by biomolecular surface is very important as they exhibit overall charge and hydrophobicity and also heterogeneous distribution of these.

- Biomolecules of interest can be broadly grouped into
  (*a*) Nucleic acids → DNA, RNA, PNA
  (*b*) Proteins → Antibodies, enzymes, receptors
  (*c*) Small molecules → Peptids, Metabolites
  (*d*) Other biomolecules → Carbohydrates, lipids.
- Oligonucleotides shows more ordered and predictable patterns, with regularly spaced –vely charged PO_4 groups and hydrophobic base pairing regions.
  In contrast, proteins are characterized by both heterogeneous and irregular regions of +ve charge, –ve charge and hydrophobility.
- Mechanisms of immobilization can be divided into two major categories:
  (1) Adsorption—Relies on non-covalent interactions, mainly electrostatic, vander Waals and dehydration of hydrophobic interfaces.
  (2) Covalent—Binding of specific functional groups on the biomolecule to functionalized surfaces.

Adsorption—purely physical nature. Therefore displays varying levels of revesibility.

Covalent—Formation of essentially irreverssible chemical bonds between biomolecule and surface. A variety of side groups namely amino, carboxy, hydroxy and thiol groups involved in covalent bonding.

DNA Immobilization on Surfaces

- DNA immobilization at solid/liquid interface is not fully understood.
- Neverthless at neutral pH, DNA molecules are –vely charged, and back bone of DNA through phosphates expected to bind strongly to a +vely charged surface, bases faces upward position.
- +vely charged surfaces like APTES—Amino propyl triethoxy silane and poly-L-lysine coated glass—commonly used in DNA hybridization.
- DNA binds to nitrocellulose and nylon supports through hydrophobic interactions.

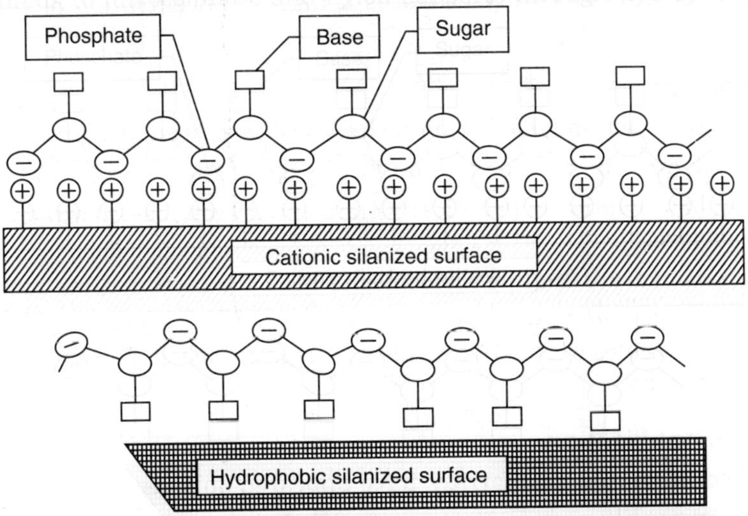

Fig. 1.13 Combinations of the DNA/oligonucleotide—surface complex on cationic, Hydrophobic surfaces (Reprinted with permission from (7). Copyright on line Elsevier Science)

The efficiency hybridization depend on

(*i*) Surface characteristics

(*ii*) Surface density of probe molecules

(*iii*) Probe orientation on the surface

(*iv*) Factors controlling transport of target molecules to the surface.

Protein Immobilization on Surfaces

- Fundamentals of protein attachment on surfaces have been widely discussed and reviewed.
- Neverthless any prediction regarding attachment is difficult because of their extreme diversity and complexity of proteins.
- However certain strategies used can be broadly classified based on adsorptive or covalent binding mechanisms.

Random Adsorptive Attachment

- As proteins are charged biomolecules, electrostatic interactions can be efficient for immobilization, but this is of a non-permanent nature and can be influenced by pH and ionic activity.
- Hydrophobic interactions—often stronger and less reversible than electrostatic, but can result in loss of functional activity due to partial denaturation.

  Therefore a reasonable approach would be to explore combinatorially level adsorption versus descriptors of surface, solution and protein characteristics.
- This method is often used for microarray purposes.

Random Covalent Attachment

- Covalent binding is quite commonly used in biochip fabrication.
- Proteins present a variety of functional groups, including amino-carboxyl-,hydroxyl- and thiol-, which can readily be used for covalent binding to surfaces with complementary chemical group.
- Covalent binding generally produces a higher concentration of protein than does adsorption.

Oriented Attachment

A variety of oriented immobilization techniques have been attempted like

(*a*) Class of Ab binding proteins to bind Fc portion of Abs.

(*b*) Terminal biotinylation of genetically engineered proteins—to attach for streptavidin coated surface.

(*c*) Use of carbohydrate binding molecules to bind the carbohydrate moieties of Abs.

- When 'Fab' fragments were specifically oriented in a dense monolayer, 90% of the adsorbed molecules were active, while randomly attached. Fab fragments were packed at lower density.

Carbohydrate Immobilization

- Carbohydrate based microarrays have appeared only recently.
- Carbohydrates like proteins are structurally heterogeneous and require preservation of molecular conformation, 3-D structure and topological configuration on a chip.

- Can be (larger) easily immobilized easily on relatively hydrophobic like nitrocellulose or treated polystyrene surfaces.
- For smaller carbohydrates synthetic glycoconjugates have been employed.

Immobilization of Cells on Surfaces

- More complex to fix compared to individual molecules.
- Depends on complexity of cell membrane—membrane proteins.

MICROARRAYS

- A microarray is a tool for analyzing gene expression that consists of a small membrane on glass slide containing samples of many genes arranged in a regular pattern.
- What exactly is a DNA microarray ?

  DNA microarrays are small, solid supports onto which the sequences from thousands of different genes are immobilized or attached, at fixed locations.
- Supports are usually glass microscope slide, but can also be silicon chips or nylon membranes.
- DNA is presented, spotted, or actually synthesized directly onto the support.
- "Array"—means—"To place on an orderly arrangement."
- It is important that the gene sequences in a microarray are attached to their support in an orderly or fixed way, because a researcher uses the location of each spot in the array to identify a particular gene sequence.
- The spots themselves can be DNA, cDNA or oligonucleotides. An oligonucleotide, as oligo as it is commonly called, is a short fragment of a ssDNA that is typically 5–50 NTs long.
- Why are Microarrays important ?
- They are significant because they contain very large number of genes and also because of their small size.
- Microarrays are therefore useful when one wants to survey a long number of genes quickly or when the sample to be studied is small.
- Microarrays may be used to array gene expression within a single sample or to compare gene expression in two different cell types or tissue sample like healthy and diseased tissue.
- In future, this technology may allow to interprete the functions of new genes based on similarities in expression patterns with those of known genes.

They may be useful in studying gene families, reveal new patterns, and uncover entirely new catagories of genes.

Use of microarrays may speed up the identification of genes involved in the development of various diseases by enabling scientists to examine a much larger number of genes.

Technology can reveal how multiple gene products work together to produce physical and chemical responses to both static and changing cellular needs.

DNA Microarray Experiment

1. Prepare your DNA chip using your chosen target DNAs.
2. Generate a hybridization solution containing a mixture of fluoroscently labeled cDNAs. (Different tags Red and Green are used so that samples can be differentiated in subsequent steps.)
3. Incubate your hybridization mixture contaning fluorescently labeled cDNAs with your DNA chip.
4. Detect bound cDNA using laser technology and store data in a computer.
5. Analyse data using computational methods.

Colors of Microarray

Green represents control DNA, where either DNA or cDNA derived from normal tissue is hybridized to the target DNA.

Red represents sample DNA, where either DNA or cDNA is derived from diseased tissue hybridized to the target DNA.

Yellow represents a combination of control and sample DNA, where both hybridized equally to the target DNA.

Black represents areas where neither the control nor sample DNA is hybridized to the target DNA.

Each spot on an array is associated with a particular gene. Each colour in an array represents either healthy (control) or diseased (sample) tissue. Depending on the type of array used, the location and intensity of a colour will tell us whether the gene, or mutation, is present in either the control and/or sample DNA. It will also provide an estimate of the expression level of the gene(s) in the sample and control DNA.

Microarray Applications

Type	Application
1. Comparative Genomic hybridization	Tumor classification, risk assessment and progressive prediction.
2. Expression analysis	Drug development, drug response and therapy development.
3. Mutation/Polymorphism analysis	Drug development, therapy development and tracking disease progression.

2

Quantum Dots

- A quantum dot is a nanoscale semiconductor crystal, only a few atoms across.
- Quantum dots can be synthetically manufactured in the lab.
- They are composed of various materials, such as lead sulfide, zinc sulfide, cadmium selenide and Indium phosphide.

  Most common one is cadmium selenide (CdSe) core surrounded by a zinc sulfide (ZnS).

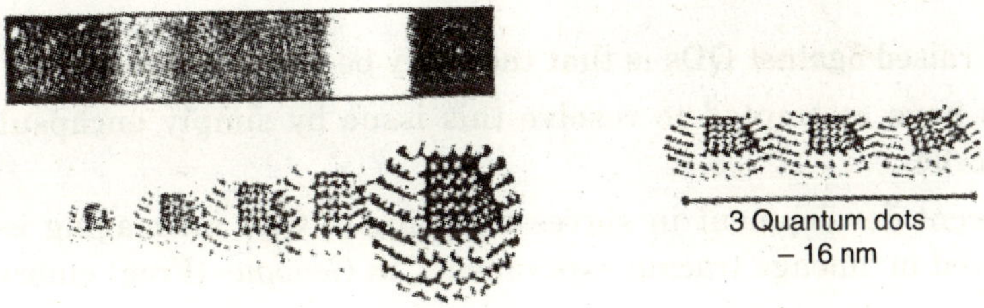

Fig. 2.1 Size dependent colour

- Most important characteristic of quantum dots is their ability to fluoresce brightly.

  This phenomenon is dependent on their size and composition, they emit a particular wavelength, or colour, of light after irradiated by an ultraviolet light, which excites their electrons.
- When hit with light, smaller quantum dots fluoresce at shorter wavelengths, such as blue, which larger dots fluoresce at longer wavelength such as red.
- By varying the size of the dots during production, scientists have boon ablo to create a dizzying array of colours of dots—from red, yellow, green, blue, violet and dozens of more colours in between.

Putting Quantum Dots to Work

Quantum dots are very neat tiny things. They can be put into use in many ways.

"These tiny nanocrystals indeed have the power to revolutionize the study of cell biology, the treatment of cancer; computing and laser technology."

■ QUANTUM DOTS AND CELLULAR IMAGING

A primary method scientists use to learn about the structure and function of cells is by tracking the production and movement of proteins within the cell.

Scientists typically do this by tagging desired protein with fluorescent marker like green fluorescent protein—**GFP** or **Rhodamine**.

However, there is disadvantage associated with conventional fluorescent tags. The excitation spectra from these interfere with each other and prevent simultaneous viewing of all these markers.

Quantum dots on the other hand, fluoresce at a unique wave length based on their size, yet can all be excited by the same wavelength of light and do not interfere with each other.

In comparison to conventional fluorophores, quantum dots are up to 100 times brighter and 100 times more resistant to fading.

Hydrophobicity

Quantum dots are hydrophobic nature, therefore scientists are using hydrophobic molecules to mark the surface and then covalently attach proteins to the dot surface, there by making the dots bind covalently and specifically to the target proteins and other molecules.

Toxicity

One issue raised against QDs is that they may be toxic to living cells.

Scientists have attempted to resolve this issue by simply encapsulating QDs in a phospholipid micelle.

- The recent development in successful using of QDs in imaging is that they have been used in 'lineage tracing experiments' in *Genopus* (Frog) embryo.
- Previously such studies were impossible, as dyes and other fluorophores fade too quickly and are not equally distributed during division.

Cancer Imaging with QDs

Cancer is one of the leading causes of death across the globe. Though there a number of treatment options are available like chemotherapy, radiation therapy, surgery—the success of these treatments is basic on the elimination of all cancerous cells from the body.

Being able to image the spread of cancerous cells in the body would be a great boon to both physicians and cancer patients. It will help the surgeons to plan precisely. QDs are well suited for this job.

Scientists have been able to bioconjugate the QDs so that they bind specifically to a desired protein/cell surface receptor/molecule.

This ability to detect the location and spread of cancer through QDs could revolutionize cancer treatment.

Scientists are not only looking at using QDs for imaging but also it is hypothesized to develop therapies with QDs.

QDs could be conjugated to both a molecule selective for the cancer cell and a chemotherapy drug molecule.

QD will be engineered to release drug only, when bound to a cancer cell. Therefore will minimize side effects by not harming healthy cells.

Building a Better Laser

- QDs can revolutionize laser technology.
- Laser is essentially a beam of directional, coherent, monochromatic light produced by stimulated emission.
- Lasers are employed in many technologies.
- CD and DVD players. Laser pointers, Spectrometry Equipment—drilling, cutting and welding machinery.
- Telecommunications technology to holography technologies.
- Quantum dot based laser can be advantageous as they form dense array of equal sized QDs in the lasers active region.
- Superior yield from QDs—would make for more efficient compact lasers.

Other Applications

- Researchers are hoping that QDs could eventually provide an efficient lighting for general use—home, office, neighbourhood and lamp.

Here, a light-emitting diode (LED) or other source would shine on QDs, which would then light up.

By mixing different sizes of QDs—one can generate fluorescent light.

It is something similar to general fluorescent tube lights without bulkyness and also less heat.

- Flat-panel TV displays using QD LEDs may provide vibrant colours than current flat-panel displays based Liquid Crystal Display (LCD) technology.

Products

1. Paul Alivisatos's Quantum Dot Corporation (QDC)—QD surrounded by an shell that amplifies its optical properties. Q Dot can attach to specific penetrate and light it up from inside.
2. QDC joined with Panasonic and Sumitomo Biosciences Corporation develop advanced optical and image processing. Expect to generate Research of more than 100 million per year by 2007.
3. Researchers at Emory University and Cambridge have developed QD identify tumors in mice.

■ HYBRID BIOLOGICAL-INORGANIC DEVICES

Bionanocomposites

- Definition of nanocomposite material has broadened significantly. It encompasses variety of systems such as one-dimensional, two-dimensional, three-dimensional and amorphous materials, made of distinctly dissimilar components and mixed at the nanometer scale.

- General class of nanocomposite organic/inorganic materials is a fast growing area of research.

There has been an increased focus to obtain viable nanoscale structures *via* innovative synthetic approaches.

The property of nanocomposite materials depend not only on the properties of their individual parents but also on their morphology and interfacial characteristics.

Research in the field may yield new materials with novel properties. There is also the possibility of new properties which are unknown in the parent constituent materials.

- At the center for bionanocomposites research, focus is on biological-inorganic interfaces, then design, synthesis and characterization of novel amalgams that fuse biological and inorganic materials.
- The integration of **"Soft"** biological and organic molecular assemblies with **"hard"** inorganic nano-architectures is of special interest. This is because the opportunity to combine normally disparate chemical and physical properties within a single system.
- This research will lead to the creation of entirely new classes of materials with tailored functionalities under the soft biological components.

Example: Nanoscale inorganic materials show robustness in catalytic and electronic reactivities, and this can be combined with responsiveness and molecular selectivity of biological molecules.

Such nanostructured, bio-inorganic amalgams can be applied in a broad range of areas like

- Chemical catalysis
- Sensors
- Information storage
- Computing
- Artificial vision
- Biological intervention
- Environmental sensing and remediation.

Conceptual Approach

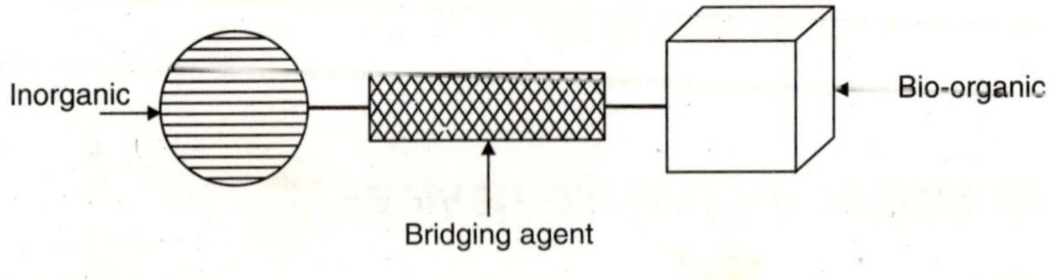

Fig. 2.2

Activities

- Bio-inorganic composites
- Nanostructured carbon-based materials
- Biosensors and devices
- Manipulation of biomolecules by using nanoparticles
- Complex fluids and membranes
- Nanocatalysis
- Chemical assembly—directed/templated self assembly.
- Nanoparticle synthesis of metals, semiconductors and insulators.
- Nanomaterials for energy.
- X-ray, electron, neutron and photonic characterization.
- Theory and modeling

Some of the outcome may be like

- Catalysts that combine the reactivity of nanostructured inorganics with the selectivity and complexity of biological molecules.
- Hybrid nanocomposites that integrate biology with electronics.
- Functional nanostructured materials that facilitate biological energy storage and transduction.
- Actuators, transducers, and power sources that serve as active elements for supramolecular machines.

Dendrimer Nanocomposites in Medicine

- Dendrimer based nanocomposites DNCs are novel organic-inorganic hybrid nanoparticles.
- They are synthesized by dispersing very small inorganic domains in nanoscopic size polymeric networks.
- They display unique material properties in addition to properties represented by their individual components.
- DNC have a great potential in biomedical applications due to their controlled composition, predetermined size, shape and optional surface functionalities.
- They may be used as drug delivery vehicles, with an active drug or radioactive isotope. Radioactive dendrimer nanocomposities can be delivered directly to the tumor by injecting nanoparticle solutions directly into the tumor microvasculature.
- DNCs can be equipped with targeting moieties that can detect cancerous cells anywhere in the body.

Poss Based Nanocomposites

- New type of hybrid based on silsesquioxane cage structure.
- Regarded as **'Molecular Silica',** smallest form of silica.
- Developed by Hybrid plastics (USA) and is called as 'Polyhedral Oligomeric Silsesquioxane (POSS) Nanotechnology.
- The particle diameter of the POSS molecules is in the range of 0.7–30Å.

- The nature of the functional group determines compatibility with polymer matrix.
- The R group can be simple alkyl, cycloalkyl or aryl or reactive polymerisable groups such as acrylic, α–olefin, styrene, epoxide, carboxylic acid, isocyanate, amine, alcohol and silane.

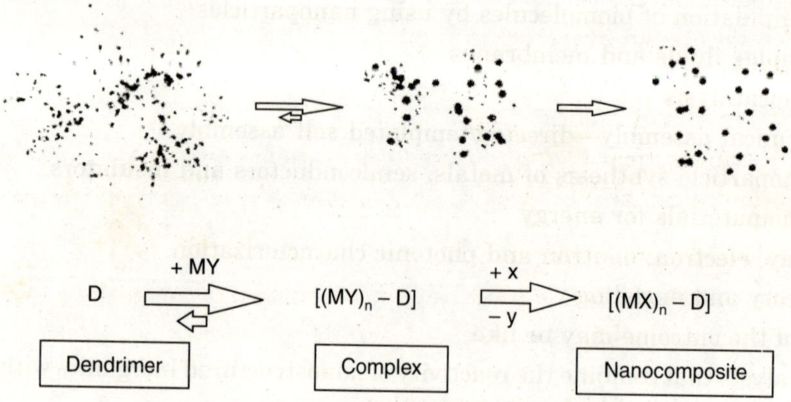

Fig. 2.3 Dendrimer complex nanocomposite

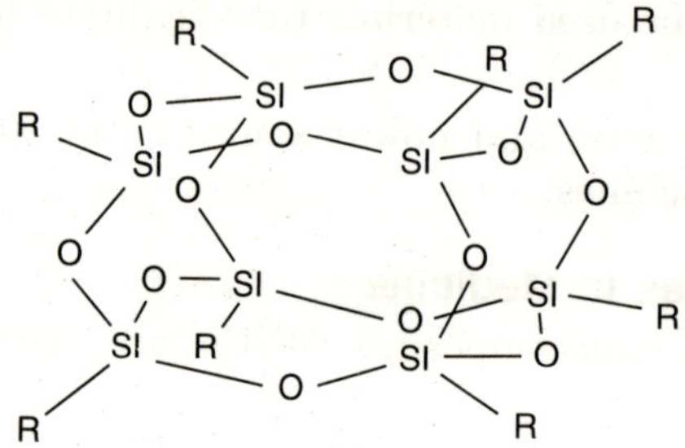

Fig. 2.4 Representation of a polyhedral oligomeric silsesquioxane or POSS

Ceramics

- Biomolecule/inorganic interactions can be used to produce ceramics with increased toughness.
- There have been increase in our understanding by the way in which an organic substance (usually a protein/peptide/lipid) interacts with an inorganic phase (like calcium carbonate or hydroxyapatite) and has led to the synthesis of many biocomposite materials.

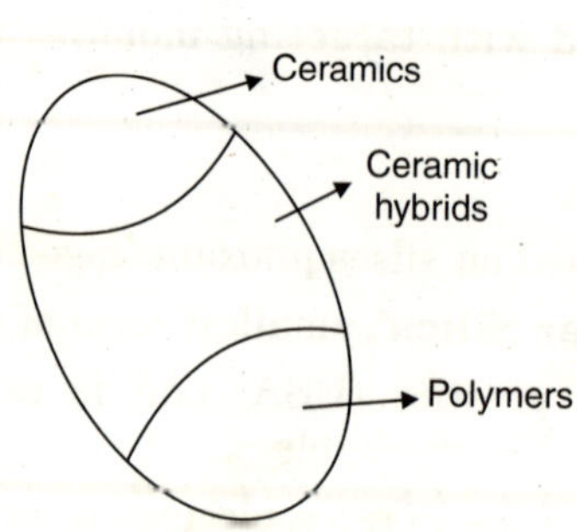

Fig. 2.5

- The structure and porosity of the inorganic phase can be controlled by templating with an organic surfactant, vesicular arrays or liquid crystalline materials.
- Micelle-templated synthesis can produce ceramics with 20–100 Å pore dimensions.

These pores can be used as

(*a*) Catalysts and absorbents.

(*b*) Gas/liquid separations.

(*c*) Thermal and acoustic insulation.

(*d*) Pharmaceutical and biochemical separations (due to their selectivity).

- Bioceramics can also be made that are more compatible with teeth and bone.

Packaging Material

- Another interesting example of an organic/inorganic composite.
- Can replace polystyrene, which otherwise widely employed for packing fast food products.
- Material is composed of potatostarch and calcium carbonate.
- Combination enjoys good thermal insulation, light weigh and biodegradable.

3

Bio MEMS and Nanotechnology

- **MEMS** stand for Microelectromechanical systems.
- The microelectronics, microfabrication and micromachining technologies are collectively known as **MEMS.**
- MEMS application to the field of Biomedical sciences, has now become a new field of research unto itself and is now known as **Bio MEMS.**
- Technology is originally based on the same technology that has been used to make computer chips ever more powerful and less expensive.
- Bio MEMS applies the above technical information and competency into diverse areas in biomedical research and clinical medicine.
- Bio MEMS is an enabling technology for improved medical diagnosis, and therapies at low cost.
- The inherent characteristics of Bio MEMS promise the production of miniature, smart and low cost biomedical devices that could revolutionize biomedical investigation and clinical practice.
- Bio MEMS and associated nanotechnology can be applied for clinical applications such as surgical instruments, tissue repair, artificial organs, diagnostic tools and drug delivery systems.
- Bio MEMS is increasingly focusing on micro/nanometer—sized features for tissue engineering, protein analysis, assays and cell interrogation.
- Miniature versions are being developed for drug delivery system transducers for ultrasound images and *in situ* telemetrically monitored pressure/temperature sensors for minimally-invasive surgery/follow up.

There has been successful in developing certain instruments like:

1. Magnetic flow cell sorting machines for various diagnostic and therapeutic applications, such as rapid screening for cancer cells in blood, human peripheral lymphocytes, cultured cell lines and samples like bonemarrow donated by patients. Bar-Ilan University in Israel.
2. Cell tracking velocimetry—can analyse individual cell velocities of hundreds of cells at a time, yielding data about the population average and dispersion, can sort some 10 million cells/second. Developed in Ohio State University, USA.

BIOSIGNAL TRANSDUCTION MECHANISMS

- Signal Transducing Receptor Classes
- Receptor Tyrosine Kinases
- Non-Receptor Tyrosine Kinases
- Receptor Serine Threonine Kinases
- Non-Receptor Serine/Threonine Kinases
- Phospholipids and Phospholipases
- G-Protein Coupled Receptors
- G-Protein Regulators
- Intracellular Hormone Receptors
- Phosphatases

Mechanisms of Signal Transduction

Signal transduction at the cellular level refers to the movement of signals from outside the cell to inside. The movement of signals can be simple, like that associated with receptor molecules of the acetylcholine class receptors that constitute channels which, upon ligand interaction, allow signals to be passed in the form of small ion movement, either into or out of the cell. These ion movements result in changes in the electrical potential of the cells that, in turn, propagates the signal along the cell. More complex signal transduction involves the coupling of ligand-receptor interactions to many intracellular events. These events include phosphorylations by tyrosine kinases and/or serine/threonine kinases. Protein phosphorylations change enzyme activities and protein conformations. The eventual outcome is an alteration in cellular activity and changes in the program of genes expressed within the responding cells.

Classifications of Signal Transducing Receptors

Signal transducing receptors are of three general classes:

1. Receptors that penetrate the plasma membrane and have intrinsic enzymatic activity
   Receptors that have intrinsic enzymatic activities include those that are **tyrosine kinases** (*e.g.*, PDGF, insulin, EGF and FGF receptors), **tyrosine phosphatases** (*e.g.*, CD45 [*cluster determinant-45*] protein of T-cells and macrophages), **guanylate cyclases** (*e.g.*, natriuretic peptide receptors) and **serine/threonine kinases** (*e.g.*, activin and TGF-□ receptors). Receptors with intrinsic tyrosine kinase activity are capable of autophosphorylation as well as phosphorylation of other substrates.

   Additionally, several families of receptors lack intrinsic enzyme activity, yet are coupled to intracellular tyrosine kinases by direct protein-protein interactions.
2. Receptors that are coupled, inside the cell, to GTP-binding and hydrolyzing proteins (termed **G-proteins**). Receptors of the class that interact with G-proteins all have a

structure that is characterized by 7 transmembrane spanning domains. These receptors are termed *serpentine* receptors. Examples of this class are the adrenergic receptors, odorant receptors, and certain hormone receptors (*e.g.*, glucagon, angiotensin, vasopressin and bradykinin).

3. Receptors that are found intracellularly and upon ligand binding migrate to the nucleus where the ligand-receptor complex directly affects gene transcription.

Receptor Tyrosine Kinases (RTKs)

The proteins encoding RTKs contain four major domains:

- An extracellular ligand binding domain.
- An intracellular tyrosine kinase domain.
- An intracellular regulatory domain.
- A transmembrane domain.

The amino acid sequences of the tyrosine kinase domains of RTKs are highly conserved with those of cAMP-dependent protein kinase (**PKA**) within the ATP binding and substrate binding regions. Some RTKs have an insertion of non-kinase domain amino acids into the kinases domain termed the kinase insert. RTK proteins are classified into families based upon structural features in their extracellular portions (as well as the presence or absence of a kinase insert) which include the cysteine rich domains, immunoglobulin-like domains, leucine-rich domains, Kringle domains, cadherin domains, fibronectin type III repeats, discoindin I-like domains, acidic domains, and EGF-like domains. Based upon the presence of these various extracellular domains the RTKs have been sub-divided into at least 14 different families.

Characteristics of the Common Classes of RTKs

Class	Examples	Structural Features of Class
I	EGF receptor, NEU/HER2, HER3	cysteine-rich sequences
II	insulin receptor, IGF-1 receptor	cysteine-rich sequences; characterized by disulfide-linked heterotetramers
III	PDGF receptors, *c*-Kit	contain 5 immunoglobulin-like domains; contain the kinase insert
IV	FGF receptors	contain 3 immunoglobulin-like domains as well as the kinase insert; acidic domain
V	vascular endothelial cell growth factor (VEGF) receptor	contain 7 immunoglobulin-like domains as well as the kinase insert domain
VI	hepatocyte growth factor (HGF) and scatter factor (SC) receptors	heterodimeric like the class II receptors except that one of the two protein subunits is completely extracellular. The HGF receptor is a proto-oncogene that was originally identified as the Met oncogene
VII	neurotrophin receptor family (trkA, trkB, trkC) and NGF receptor	contain no or few cysteine-rich domains; NGFR has leucine rich domain

Many receptors that have intrinsic tyrosine kinase activity as well as the tyrosine kinases that are associated with cell surface receptors contain tyrosines residues, that upon phosphorylation, interact with other proteins of the signaling cascade. These other proteins contain a domain of amino acid sequences that are homologous to a domain first identified in the c-Src proto-oncogene (*c*-designates the cellular form of proto-onogenes that were first identified in transforming retrovirus). These domains are termed **SH2 domains** (Src homology domain 2). Another conserved protein-protein interaction domain identified in many signal transduction proteins is related to a third domain in c-Src identified as the **SH3 domain.**

The interactions of SH2 domain containing proteins with RTKs or receptor associated tyrosine kinases leads to tyrosine phosphorylation of the SH2 containing proteins. The result of the phosphorylation of SH2 containing proteins that have enzymatic activity is an alteration (either positively or negatively) in that activity. Several SH2 containing proteins that have intrinsic enzymatic activity include **phospholipase C-□ (PLC-□),** the proto-oncogene c-Ras associated GTPase activating protein **(rasGAP), phosphatidylinositol-3-kinase (PI-3K), protein phosphatase-1C (PTP1C),** as well as members of the Src family of **protein tyrosine kinases (PTKs).**

Non-Receptor Protein Tyrosine Kinases (PTKs)

There are numerous intracellular PTKs that are responsible for phosphorylating a variety of intracellular proteins on tyrosine residues following activation of cellular growth and proliferation signals. There is now recognized two distinct families of non-receptor PTKs. The archetypal PTK family is related to the Src protein. The Src protein is a tyrosine kinase first identified as the transforming protein in Rous sarcoma virus. Subsequently, a cellular homolog was identified as **c-Src.** Numerous proto-oncogenes were identified as the transforming proteins carried by retroviruses. The second family is related to the **Janus kinase (Jak).** Most of the proteins of both families of non-receptor PTKs couple to cellular receptors that lack enzymatic activity themselves. This class of receptors includes all of the cytokine receptors (*e.g.,* the interleukin-2 (IL-2) receptor) as well as the CD4 and CD8 cell surface glycoproteins of T cells and the T cell antigen receptor (TCR). This mode of coupling receptors to intracellular PTKs suggests a split form of RTK.

Another example of receptor-signaling through protein interaction involves the insulin receptor (IR). This receptor has intrinsic tyrosine kinase activity but does not directly interact, following autophosphorylation, with enzymatically active proteins containing SH2 domains (*e.g.,* PI-3K or PLC-□). Instead, the principal IR substrate is a protein termed IRS-1, IRS-1 contains several motifs that resemble SH2 binding consensus sites for the catalytically active subunit of PI-3K. These domains allow complexes to form between IRS-1 and PI-3K. This model suggests that IRS-1 acts as a **docking** or **adapter protein** to couple the IR to SH2 containing signaling proteins.

Additional adapter proteins have been identified, the most commonly occurring being a protein termed **growth factor receptor-binding protein 2, Grb2.**

An example of an alteration in receptor activity in response to association with an intracellular PTK is the nicotinic acetylcholine receptor (AChR). These receptors comprise an ion channel consisting of four distinct subunits (□, □, □, and □).

The □, □, and □ subunits are tyrosine phosphorylated in response to acetylcholine binding which leads to an increase in the rate of desensitization to acetylcholine.

Receptor Serine/Threonine Kinases (RSTKs)

The receptors for the TGF-□ superfamily of ligands have intrinsic serine/threonine kinase activity. A more complete description of the TGF-□ signaling cascade can be found in the Signaling by Wnts and the TGFs-b/BMP Families page.

There are more than 30 multifunctional proteins of the TGF-□ superfamily which also includes the activins, inhibins and the bone morphogenetic proteins (BMPs). This superfamily of proteins can induce and/or inhibit cellular proliferation or differentiation and regulate migration and adhesion of various cell types. The signaling pathways utilized by the TGF-□, activin and BMP receptors are different than those for receptors with intrinsic tyrosine kinase activity or that associate with intracellular tyrosine kinases.

At least 17 RSTKs have been isolated and can be divided into 2 subfamilies identified as the type I and type II receptors. Ligands first bind to the type II receptors which then leads to interaction with the type I receptors. When the complex between ligand and the 2 receptor subtypes forms, the type II receptor phosphorylates the type I receptor leading to initiation of the signaling cascade.

One predominant effect of TGF-□ is regulation of progression through the cell cycle. One nuclear protein involved in the responses of cells to TGF-□ is the proto-oncogene, c-Myc ("mick") which directly affects the expression of genes harboring Myc-binding elements.

Non-Receptor Serine/ Threonine Kinases

There are several serine/threonine kinases that function in signal transduction pathways. The two more commonly known are **cAMP-dependent protein kinase (PKA)** and **protein kinase C (PKC).** Additional serine/threonine kinases important for signal transduction are the **mitogen activated protein kinases (MAP kinases).**

Protein Kinase C (PKC)

PKC was originally identified as a serine/threonine kinase that was maximally active in the presence of diacylglycerols (DAG) and calcium ion. It is now known that there are at least ten proteins of the PKC family. Each of these enzymes exhibits specific patterns of tissue expression and activation by lipid and calcium. PKCs are involved in the signal transduction pathways initiated by certain hormones, growth factors and neurotransmitters. The phosphorylation of various proteins, by PKC, can lead to either increased or decreased activity. Of particular importance is the phosphorylation of the EGF receptor by PKC which down-regulates the tyrosine kinase activity of the receptor. This effectively limits the length of the cellular responses initiated through the EGF receptor.

MAP Kinases

MAP kinases were identified by virtue of their activation in response to growth factor stimulation of cells in culture, hence the name **mitogen activated protein kinases.** MAP kinases are also called ERKs for **evtracellular-signal regulated kinases.** On the basis of *in vitro* substrates the MAP kinases have been variously called microtubule associated protein-2 kinase (MAP-2 kinase), myelin basic protein kinase (MBP kinase), ribosomal S6 protein kinase (RSK-kinase: *i.e.,* a kinase that phosphorylates a kinase) and EGF receptor threonine kinase (ERT kinase). All of these proteins have similar biochemical properties, immuno-crossreactivities, amino acid sequence and ability to *in vitro* phosphorylate similar substrates.

Maximal MAP kinase activity requires that both tyrosine and threonine residues are phosphorylated. This indicates that MAP kinases act as switch kinases that transmits information of increased intracellular tyrosine phosphorylation to that of serine/threonine phosphorylation. Although MAP kinase activation was first observed in response to activation of the EGE, PDGF, NGF and insulin receptors, other cellular stimuli such as T cell activation (which signals through the Lck [*lick*] tyrosine kinase), phorbol esters (that function through activation of PKC), thrombin, bombesin and bradykinin (that function through G-proteins) as well as N-methyl-D-aspartate (NMDA) receptor activation and electrical stimulation rapidly induce tyrosine phosphorylation of MAP kinases.

MAP kinases are, however, not the direct substrates for RTKs nor receptor associated tyrosine kinases but are in fact activated by an additional class of kinases termed MAP kinase kinases (MAPK kinases) and MAPK kinase kinases (MAPKK kinases). One of the MAPK kinases has been identified as the proto-oncogenic serine/threonine kinase, Raf. Ultimate targets of the MAP kinases are several transcriptional regulators *e.g.*, serum response factor (SRF), and the proto-oncogenes Fos, Myc and Jun as well as members of the steroid/thyroid hormone receptor super family of proteins.

Phospholipases and Phospholipids in Signal Transduction

Phospholipases and phospholipids are involved in the processes of transmitting ligand-receptor induced signals from the plasma membrane to intracellular proteins. The primary protein affected by the activation of phospholipases is PKC which is maximally active in the presence of calcium ion and DAG. The generation of DAG occurs in response to agonist activation of various phospholipases. The principal mediators of PKC activity are receptors coupled to activation of phospholipase C-□ (PLC-□). PLC-□ contains SH2 domains that allow it to interact with tyrosine phosphorylated RTKs. This allows PLC-□ to be intimately associated with the signal transduction complexes of the membrane as well as membrane phospholipids that are its substrates. Activation of PLC-□ leads primarily to the hydrolysis of membrane phosphatidylinositol bisphosphate (PIP_2) leading to an increase in intracellular DAG and inositol trisphosphate (IP_3). The released IP_3 interacts with intracellular membrane receptors leading to an increased release of stored calcium ions. Together, the increased DAG and intracellular free calcium ion concentrations lead to increased activity of PKC. Recent evidence indicates that phospholipases D and A_2 (PLD and PLA_2) also are involved in the sustained activation of PKC through their hydrolysis of membrane phosphatidylcholine (PC). PLD action on PC leads to the release of phosphatidic acid which in turn is converted to DAG by a specific phosphatidic acid phosphomonoesterase. PLA_2 hydrolyzes PC to yield free fatty acids and lysoPC both of which have been shown to potentiate the DAG mediated activation of PKC. Of medical significance is the ability of **phorbol ester tumor promoters** to activate PKC directly. This leads to elevated and unregulated activation of PKC and the consequent disruption in normal cellular growth and proliferation control leading ultimately to neoplasia.

Phosphatidylinositol-3-Kinase (PI-3K)

PI-3K is tyrosine phosphorylated, and subsequently activated, by various RTKs and and receptor-assocaited PTKs. PI-3K is a heterodimeric protein containing an 85 kDa and 110 kDa subunits. The p85 subunit contains SH2 domains that interact with activated receptors or other receptor-assocaited PTKs and is itself subsequently tyrosine phosphorylated and activated. The 85 kDa subunit is non-catalytic, however, it does contain a domain homologous to GTPase activating (GAP) proteins. It is the 110 kDa subunit that is enzymatically active. PI-3K, associates with and is activated by, the PDGF, EGF, insulin, IGF-1, HGF and NGF receptors. PI-3K phosphorylates various phosphatidylinositols at the 3 position of the inositol ring. This activity generates additional substrates for PLC-□ allowing a cascade of DAG and IP3 to be generated by a single activated RTK Philadelphial chromosome (Ph^+) observed with high frequency in chronic myclogenous leukemias (CMLs) and acute lymphocytic leukemias (ALLs).

Intracellular Hormone Receptors

Hormone receptors are proteins that effectively bypass all of the signal transduction pathways described thus far by residing within the cytoplasm. Additionally, all of the hormone receptors are bi-functional. They are capable of binding hormone as well as directly activating gene transcription.

The steriod/thyroid hormone receptor superfamily (*e.g.*, glucocorticoid, vitamin D, retinoic acid and thyroid hormone receptors) is a class of proteins that reside in the cytoplasm and bind the lipophilic steroid/thyroid hormones. These hormones are capable of freely penetrating the hydrophobic plasma membrane. Upon binding ligand the hormone-receptor complex translocates to the nucleus and bind to specific DNA sequences termed **hormone response elements (HREs).** The binding of the complex to an HRE results in altered transcription rates of the associated gene.

Phosphatases in Signal Transduction

Substantial evidence links both tyrosine and serine/threonine phosphorylation with increased cellular growth, proliferation and differentiation. Removal of the incorporated phosphates must be a necessary event in order to turn off the proliferative signals. This suggests that phosphatases may function as anti-oncogenes or growth suppressor genes. The loss of a functional phosphatase involved in regulating growth promoting signals could lead to neoplasia. However, examples are known where dephosphorylation is required for promotion of cell growth. This is particularly true of specialized kinases that are directly involved in regulating cell cylce progression. Therefore, it is difficult to envision all phosphatases as being tumor suppressor genes.

There are two broad classes of **protein tyrosine phosphatases (PTPs).** One class are transmembrane enzymes which contain the phosphatase activity domain in the intracellular portion of the protein. The other are intracellularly localized enzymes. The first transmembrane PTP characterized was the **leukocyte common antigen protein, CD45.** This protein was shown to have homology to the intracellular PTP, PTP1B. There are at least six sub-classes of the transmembrane PTPs.

The clearest studies of a role for transmembrane PTPs in signal transduction have involved the CD45 protein. These studies have shown that CD45 is involved in the regulation of the tyrosine kinase activity of Lck in T cells. As indicated above Lck is associated with T cell antigens CD4 and CD8 generating a split-RTK involved in T cell activation. It is suspected that CD45 dephosphorylates a regulatory tyrosine phosphorylation site in the C-terminus of Lck, thereby, increasing the activity of Lck towards its substrate(s).

The second class of PTPs are the intracellular proteins. The C-terminal residues of most if not all intracellular PTPs are very hydrophobic and suggest these sites are or other protein tyrosine kinases.

G-Protein Coupled Receptors

There are several different classifications of receptors that couple signal transduction to G-proteins. These classes of receptor are termed **G-protein coupled receptors, GPCRs.** Well over 1000 different GPCRs have been cloned, most being orphan receptors having no as yet identified ligand. Three different classes of GPCR are reviewed:

1. GPCRs that modulate adenylate cyclase activity. One class of adenylate cyclase modulating receptors activate the enzyme leading to the production of cAMP as the second messenger. Receptors of this class include the □-adrenergic, glucagon and odorant molecule receptors. Increases in the production of cAMP leads to an increase

in the activity of PKA in the case of □-adrenergic and glucagon receptors. In the case of odorant molecule receptors the increase in cAMP leads to the activation of ion channels. In contrast to increased adenylate cyclase activity, the □-type adrenergic receptors are coupled to inhibitory G-proteins that repress adenylate cyclase activity upon receptor activation.

2. GPCRs that activate PLC-□ leading to hydrolysis of polyphosphoinositides (*e.g.*, PIP_2) generating the second messengers, diacylglycerol (DAG) and inositoltrisphosphate (IP_3). This class of receptors includes the angiotensin, bradykinin and vasopressin receptors.
3. A novel class of GPCRs are the photoreceptors. This class is coupled to a G-protein termed **transducin** that activates a phosphodiesterase which leads to a decrease in the level of cGMP. The drop in cGMP then results in the closing of a Na^+/Ca^+ channel leading to hyperpolarization of the cell.

G-Protein Regulators

The activity of G-proteins with respect to GTP hydrolysis is regulated by a family of proteins termed **GTTase activating proteins, GAPs**. The proto-oncogenic protein, Ras, is a G-protein involved in the genesis of numerous forms of cancer (when the protein sustains specific mutations). Of particular clinical significance is the fact that oncogenic activation of Ras occurs with higher frequency than any other gene in the development of colo-rectal cancers. Regulation of Ras GTPase activity is controlled by **rasGAP**.

There are several other GAP proteins besides rasGAP that are important in signal transduction. There are two clinically important proteins of the GAP family of proteins. One is the gene product of the neurofibromatosis type-1 (NF1) susceptibility locus. The NF1 gene is a tumor suppressor gene and the protein encoded is called **neurofibromin.** The second is the protein encoded by the BCR locus (**break point cluster region gene**). The BCR locus is rearranged in the membrane attachment domains of these proteins. One role of intracellular PTPs is in the maturation of *Xenopus* oocytes in response to hormone. Over expression of PTP-1B in oocytes resulted in a marked retardation in the rate of insulin and progesterone-induced maturation. These results suggest a role for PTP-1B in countering the signals leading to cellular activation.

The above observation as well as several others have demonstrated a link between insulin function and PTP-1B. PTP-1B directly interacts with the insulin receptor and removes the tyrosine phosphates incorporated by autophosphorylation in response to insulin binding, thereby, negatively affecting the activity of the insulin receptor. Recently, the PTP-IB gene was disrupted in mice by targeted deletion. Mice lacking a functional PTP-IB gene exhibit increased insulin sensitivity as well as resistance to obesity induced by a high fat diet. As with the transmembrane PTPs little is known about the regulation of the activity of the intracellular PTPs.

Two intracellular PTPs (PTP-IC and PTP-ID) have been shown to contain SH2 domains. These SH2 domains allow these PTPs to directly interact with tyrosine phosphorylated RTKs and PTKs, thereby, dephosphorylating tyrosines in those proteins. Following receptor stimulation of signal transduction events, the SH2 containing PTPs are directed to several of the RTKs and/or PTKs with the net effect being a termination of the signaling events by tyrosine dephosphorylation. Other phosphatases that recognize serine and/or threonine phosphorylated proteins also exist in cells. These are referred to as **protein serine phosphatases (PSPs).** At least 15 distinct PSPs have been identified. The type 2A PSPs exhibit selective substrate specificity towards PKC phosphorylated proteins; in particular serine and threonine phosphorylated receptors. Type 2A PSPs are more effective than other PSPs in dephosphorylating RSKs, proteins that are involved in signaling cascades by phosphorylating ribosomal S6 protein. However, a type 1 PSP is required to dephosphorylate S6 itself.

The type 2A PSPs have 2 subunits (a regulatory and a catalytic) both of which can associate with one of the tumor antigens of the DNA tumor virus, polyoma. Transformation by DNA tumor viruses such as polyoma appears to be mediated by the formation of a signal transduction unit consisting of a virally encoded T antigen and several host encoded proteins. Several host proteins are tyrosine kinases of the Src family. Polyoma middle T antigen also can bind to Pi-3K. The association of type 2A PSPs in these complexes may lead to dephosphorylation of regulatory serine/threonine phosphorylated sites resulting in increased signal transduction and subsequent cellular proliferation.

■ TRANSDUCERS

- We come across 3 basic instruments in biomedical instrumentation; they are:
  (*a*) Transducer—It is a device that converts energy from one form to another.
  (*b*) Sensor—Converts a physical parameter to an electric output.
  (*c*) Actuator—Converts an electric signal to a physical output.

Transducers and the Measurement of Physiological Events

- It is the task of the transducer to convert energy from one form to another, more useful form.
- Sometimes the terms sensor or detector are used to describe a device that detects a physiological event.
- In biomedical applications, the transducer converts a physiological event to an electrical signal. With the event available as an electrical signal, it is easy to obtain the advantages of modern computing and display equipment to present the desired information in the most useful form.
- Many methods are employed to convert a physiological event to an electrical signal via measuring resistance, capacitance, inductance, or the magnetic linkage between two or more coils.
- The use of piezoelectric and photoelectric transducers is also common.
- A chemically induced colour change is also a useful transduction technique.
- Practical application of principles of transduction can be applied to study of e^{-}s or ion through solids, liquids, gases or a vaccum.
- A transducer in reality is a sense organ for the electronic processing equipment.

Active Transducers—Converts one form of energy directly into another

Example. Photovoltaic cell—light energy converted to electrical energy.

Passive Transducers—Requires energy to translate changes due to the measurand.

Example. Wheatstone bridge requires ac/dc excitation signal.

Electromagnetic Flowmeters

- In 1832, Michael Faraday discovered the law governing magnetic induction.
- Faraday showed that if a conductor moved in a magnetic field, the direction of the movement of the conductor and the direction of the magnetic field must be perpendicular. Under these conditions, there is maximum induction of current.

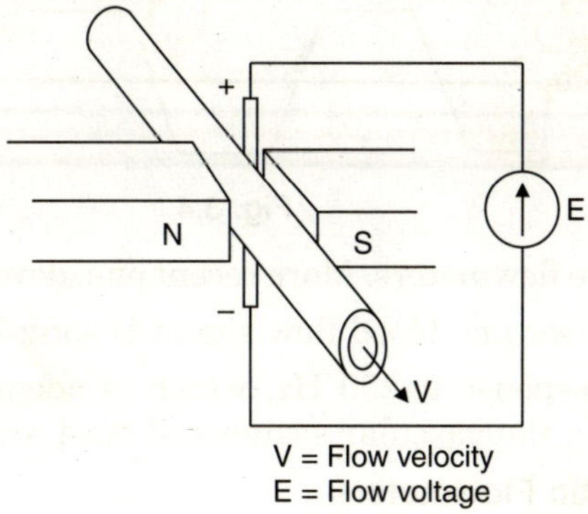

Fig. 3.1

- It took almost, a century to execute this experiment successful and it was Young *et al.*, 1920; England.
- Fabre and d'Arsonaval (1932)—exposed a femoral artery, located in a magnetic field, placed electrodes on its surface, and detected a weak flow dependent voltage—there by demonstrating that the flow signal could be detected without opening the blood vessel.

The electromagnetic flowmeter became a practical instrument with simultaneous but independent studies of Kolin and Kado of USA and Wetteror in Germany.

They placed "Nonpolarizable" electrodes on the surface of the blood vessels and used strong constant-current (dc) electromagnets to obtain high-intensity magnetic fields.

The voltage detected by the electrodes was found to be linearily proportional to blood flow velocity.

- There are many flowmeters, which differ in the way the electric signal is produced. Earlier one was, (*a*) Sine wave flowmeters was developed by Kolin in 1941.

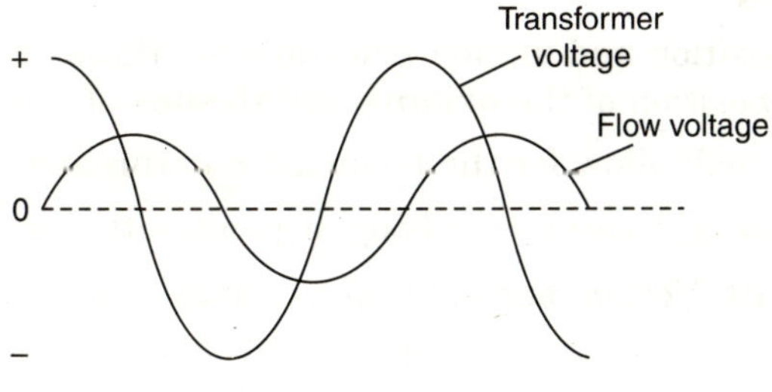

Fig. 3.2

(*b*) Square wave flowmeters was developed by Denison *et al.*, in 1953

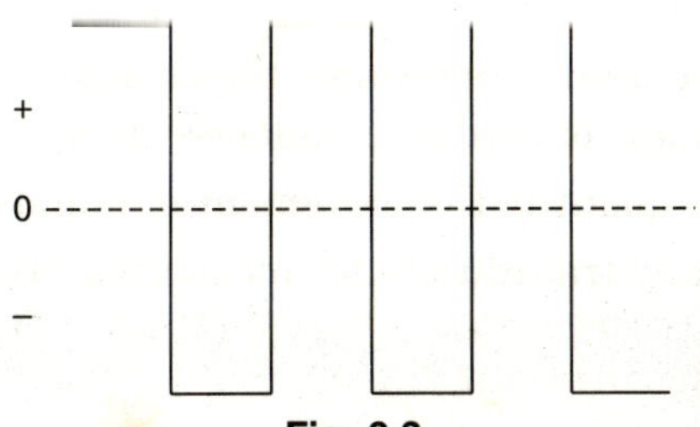

Fig. 3.3

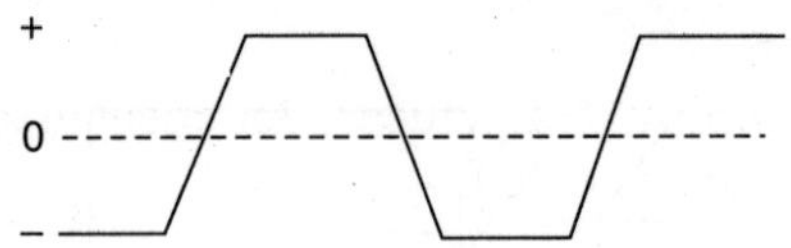

Fig. 3.4

(*c*) Trapezoidal wave flowmeters. More recent one, developed by Yanof *et al.*, in 1963.

Sampling time 0.174 m sec *i.e.,* blood flow signal is sampled 2,000 times per second.

Sinusoidal frequency response 0–250 Hz, which is adequate to show the most rapid changes in flow encountered in the vascular systems of most animals.

Case Study in Electromagnetic Flowmeters

- In animal research, it is often necessary to obtain information about the blood flow over a period of several months.
- This requirement is best met by the use of implantable flowmeters. However electromagnetic flowmeters are not suitable for implantable meters and they have to applied externally.
- The technique was developed by Kolin and Okai independently, referred to as **"Magnetorheography".**

The experiment was conducted on a frog.

- A sieve containing two electrodes were applied to flood vessel, electrodes were connected to amplifying display device and strong magnetic field was applied to this electrode assembly.
- Later Okai in 1967 applied method to the human thorax. In 1971, Okai once again successfully applied to the femoral region of an anesthelized dog.
- Electromagnetic blood flowmeters thus vary considerably in design. Some, when a periaortic or peripulmonary artery probe is used, the **ECG** is detected.

Mechanical Transducers

- Displacement, position and motion transducers—these are useful in measuring the size, shape and position of the **organs** and **tissues** of the body.
- Specifically they will allow for the following measurements to carried out
  (*a*) **Position:** Spatial location of a body or point with respect to a reference point.
  (*b*) **Displacement:** Vector representing a change in position of a body or point with respect to a reference point. Displacement may be linear or angular.
  (*c*) **Motion:** Change in position with respect to a reference system.
- Displacement transducers can be used in both direct and indirect system of measurement.

Example. They could be used to determine the change in diameter of the blood vessels and changes in volume and shape of cardiac chambers—Direct measurement.

Indirect Measurement—quantify the movement of liquids through the heart valves.

- They are basis of many transducers for measuring pressure, force and acceleration.

Potentiometric Transducers

- This is one of the simplest transducers for measuring displacement similar to the volume control on an audio electronic device.
- The resistance between two terminals on this device is related to the linear/angular displacement.

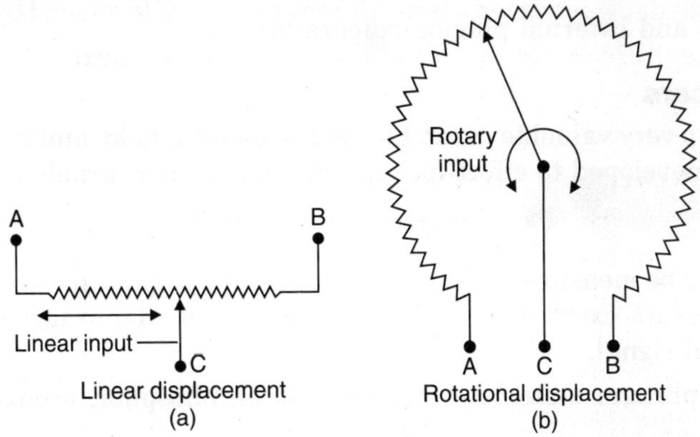

Fig. 3.5 Principle of a potentiometric displacement transducer for the measurement of (*a*) linear displacement (*b*) angular displacement

- The terminals of the potentiometer—connected the power supply either a.c. or d.c. output voltage at the wiper varies with the displacement of the subject.
- Single turn or multiturn (2–10 turns), linear, logarithmic and sectored potentiometers are available for different applications.

Piezo-electric Transducers

- The piezo-electric effect is a property of natural crystalline substances to develop electric potential along a crystallographic axis in response to the movement of change as a result of mechanical deformation.
- Therefore, piezo-electricity is pressure electricity.
- The reverse effect is also present, when an electrical field is applied, the crystal changes shape.
- On application of pressure, the charge 'Q' developed along a particular axis is given by

$$Q = dF \text{ Coulomb}$$

d = piezo-electric constant, coulombs/Newton *i.e.*, C/N and 'F' is the applied force.

- A large number of crystallographic substances are known, but the most extensively used piezo-electric crystals are quartz, ammonium dihydrogen phosphate, Rochelle salt, lithium sulfate, lead zirconate and barium titanate.

Typical values for 'd' are 2.3 C/N for quartz

140 C/N for Barium titanate

Piezo-electric materials are also available as polymeric films such as polyvinylidene fluoride—PVDF.

Medical applications

1. Numerous application in the medical instrumentation field.

2. Used in Ultrasonic scanners for imaging.

3. Used in Blood flow measurements.

4. Detection of Korotkoff sound in non-invasive blood pressure measurements.

5. In External and Internal phonocardiography.

Pressure Transducers

- Pressure is a very valuable parameter in the medical field, and therefore many devices have been developed to effect its transduction to measurable electrical signals.

Principle

The pressure to be measured is applied to a flexible diaphragm which gets deformed by the action of the pressure exerted on it. This motion of the diaphragm is then measured in terms of the electrical signal.

- Typical diaphragm materials—stainless steel, phosphor, bronze, beryllium, copper *etc*.
- Many transducers are commercially available.

1. Linear Variable Differential Transformer (LVDT) pressure transducer—LVDT pressure transducer consists of 3 parts

   (*a*) Plastic dome with two female Luer-Lok fittings

   (*b*) Stainless steel diaphragm and core assembly

   (*c*) Plastic body containing the LVDT coil assembly.

   - Transducers are available commercially in a complete array of pressure ranges.
   - LVDT pressure transducers available in two basic diaphragm and core assemblies.

First for venous and general purpose clinical measurements, with a standard size diaphragm with internal fluid volume between dome and diaphragm less than 0.5 cc.

Second with higher response characteristics for arterial pressure with fluid volume between dome and diaphragm approx 0.1 cc.

2. Strain Gauge Pressure Transducers

- Nearly all commercially available pressure monitoring systems use the strain gauge type pressure transducers for intra-arterial and IV pressure measurements.
- Transducer is based upon the changes in resistance of a wire produced due to small mechanical displacements
- A linear relation exists between the deformation and electric resistance of a suitably selected gauge (wire, foil) over a specified range.

$$\text{Gauge factor is defined as } g = \frac{\Delta R/R}{\Delta L/L} = \frac{\text{Fractional change in electric resistance}}{\text{Fractional change in length}}$$

ΔR = Incremental change in resistance due to stress

R = Resistance of an unstretched wire.

ΔL = Incremental change in length

L = Unstretched length of wire.

- Most of the pressure transducers for the direct measurement of blood pressure are of the unbonded wire strain gauge type.

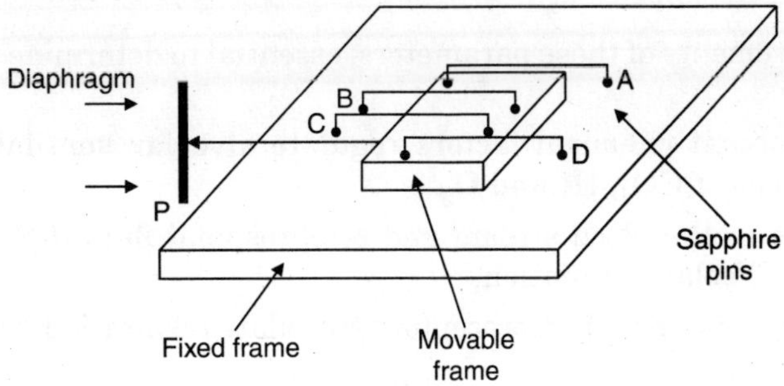

Fig. 3.6 Schematic diagram of an unbonded strain gauge pressure transducer

- Here, outer frame is fixed.
- Inner frame is connected to the diaphragm—upon which the pressure acts and is movable.
- Pressure 'P' applied in the direction shown stretches wires 'B' and 'C' and relaxes wires 'A' and 'D'.
- These wires form a four arm active bridge.
- Moving frame is mounted on springs which brings it to the central reference position, when no pressure is applied.
- These unbonded strain gauge transducers are preferred when low pressure measurements are to be made.
- Unbonded strain gauge transducers can be made sufficiently small, which are even suitable for mounting at the tip of a cardiac catheter.

Chemical Transducers

- Chemical transducers play an important role in medicine and physiology in assessing metabolism.
- There are two types of chemical transducers

  (*a*) Those that measure chemical composition of the blood, tissue and organ fluids.

  (*b*) Those that measure the composition of the respiratory gases.
- They are usually employed to measure

  (*i*) partial pressure of oxygen—pO_2

  (*ii*) partial pressure of carbondioxide—pCO_2

  (*iii*) concentration of hydrogen ions—pH.
- Complementing these transducers are those for the respiratory gases O_2, CO_2 and N_2.

Blood Gases

- Perhaps the most frequently used chemical transducers are those for measuring blood gases.

- Blood gas analysers are used to measure the pH, partial pressure of CO_2 and partial pressure of Oxygen (pO_2), of the body fluids with special reference to the human blood.
- The measurements of these parameters, essential to determine the acid-base balance in the body.
- Three important chemical factors regulate alveolar ventilation are the arterial concentrations of CO_2, H^+ and O_2.
- CO_2 tension in the blood stream and cerebrospinal fluid (**CSF**) is the major factor regulating alveolar ventilation.
- The carotid and aortic chemoreceptors stimulate respiration when O_2 tension is low.

Expected Values of Blood Gas Parameters

Parameters	—	Arterial or arterialized capillary blood	Venous plasma
pH		7.37–7.44	7.35–7.45
pCO_2	Men	34–35 mm Hg	36–50 mm Hg
	Women	31–42 mm Hg	34–50 mm Hg
pO_2	Resting adult	80–90 mm Hg	25–40 mm Hg
	Resting adult over 65 yrs	75–85 mm Hg	
Total CO_2	Men	24–30 m mol/l	26–31 m mol/l
(plasma)	Women	21–30 m mol/l	24–29 m mol/l

- Two laws need to be kept in mind before blood gas transducers are discussed:
  1. **Dalton's Law:** States that the pressure exerted by a mixture of gases is equal to the sum of the pressures that each gas would exert, it alone occupied the whole volume.
  2. **Henry's Law:** States that when a gas is exposed to a liquid, the volume 'V' that goes into a simple physical solution depends directly on the partial pressure and the solubility of the gas in the solution.

Blood pH Measurement

- The acidity or alkalinity of a solution depends on its concentration of hydrogen ions.
- pH is the negative exponent (log) of H-ion concentration pH = – log (H^+)
- Glass electrode is used for the measurement, and the potential of glass electrode can be written by Nernst equation.

Measurement of Blood pCO_2

- Blood pCO_2 is the partial pressure of CO_2 of blood taken anaerobically. It is expressed in mm Hg and is related to the percentage CO_2 as follows:

$$pCO_2 = \text{Barometric pressure} - \text{Water vapour pressure} \times \frac{CO_2}{100}$$

- Modern blood gas analyzers make use of pCO_2 electrode described by Stow *et al.*
- Basically Consists of
  → pH sensitive glass electrode—having a rubber membrane stretched over it, with a thin layer of water separating the membrane from the electode surface.

The technique is based on the fact that the dissolved CO_2 changes the pH of an aqueous solution.

↓

The CO_2 from the blood sample diffuses through the membrane to form H_2CO which dissociates into H^+ and HCO_3^- ions.

↓

The resultant change in pH is thus a function of the CO_2 concentration.

↓

The emf of generated is in linear relationship between the pH and negative logarithm of pCO_2.

Electrode construction based on Bradley and Severinghaus, 1958.

$$E = E_o - \frac{2.3036RT}{F} \Delta pH$$

E_o = Standard potential

R = Gas constant

T = Absolute temperature

F = Faraday constant

ΔpH = pH value deviation

Change of pH of one unit is 58.2 mV at 20°C and 62.2 mV at 40°C.

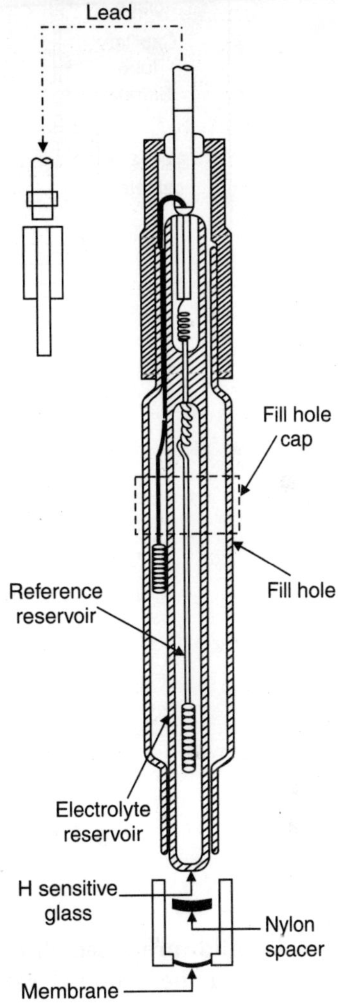

Fig. 3.7 Construction of pCO_2 electrode (Courtesy: Corning Scientific Instruments, USA)

Electrodes for Blood pH Measurement

- Several types of electrodes are described
- However the most common type is the 'Syringe electrode'—preferred for the convenience of taking small samples of blood anaerobically.
- Typically a microelectrode for clinical applications requires only 20–25 μl of capillary blood.

↓

Electrode is enclosed in a water jacket with circulating water at a constant temperature of 38°C.

↓

Water contain 1% NaCl for shielding against static interference.

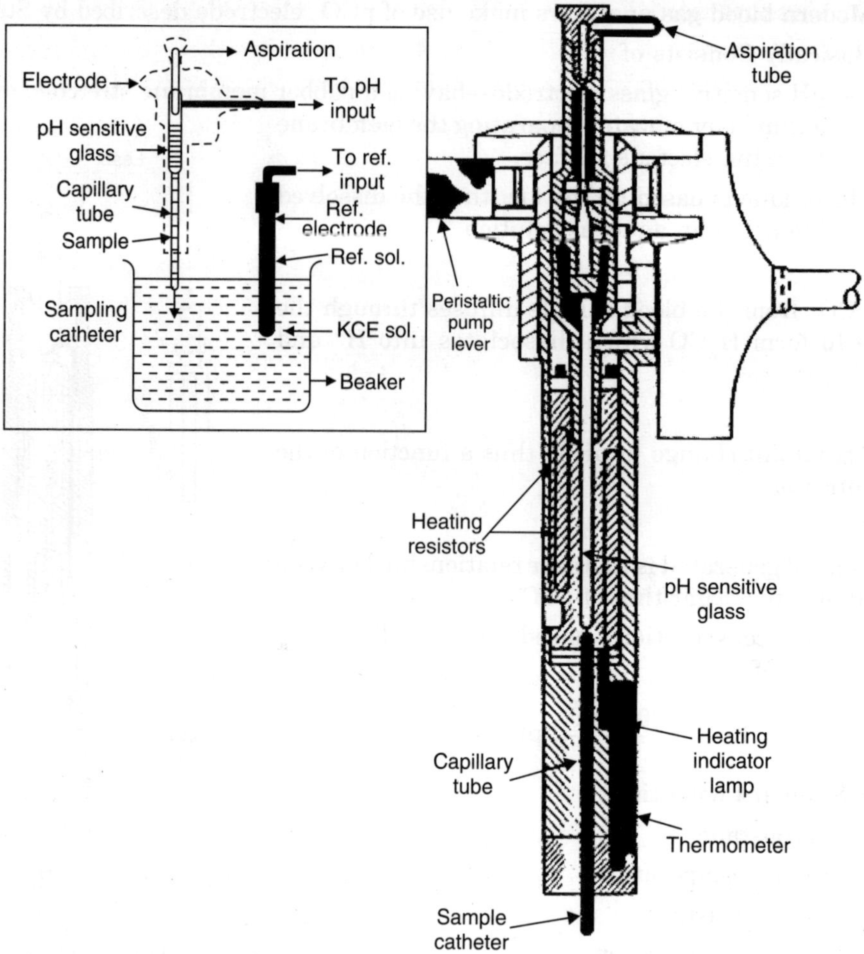

Fig. 3.8 Microcapillary electrode for measurement of blood pH (Courtesy: Corning Scientific Instruments, USA)

↓

Capillar protected with polyethylene tubing. Internal reference electrode is Silver/Silver chloride and the Calomel reference electrode—connected to a small pool of saturated KCl through a porous pin.

↓

Accuracy of 0.001 pH can be obtained with this electrode against a constant buffer.

Here water layer is replaced by a thin film of an aqueous sodium bicarbonate ($NaHCO_3$) solution.

Rubber membrane replaced by thin teflon membrane-permeable to CO_2 and not to any other ions.

↓

CO_2 from blood diffuses into the bicarbonate solution—there will be a drop in pH.

↓

pH falls by almost one pH unit for a ten-fold ↑ in the CO_2 tension of the sample, therefore pH change is linear function of the logarithm of CO_2 tension.

Blood pO_2 Measurement

- Partial pressure of O_2 in the blood/plasma indicates the extent of oxygen exchange between lungs and blood and its ability to perfuse the body tissues $\bar{c}$ oxygen.
- pO_2 is usually measured with a polarographic electrode.
- Electrode construction:
  - → Most modern blood gas analyzers utilize an oxygen electrode described by Clark, 1956
  - → Electrode consists of a platinum cathode
  - → a silver/silver chloride anode
  - → in an electrolyte filling solution and a polypropylene membrane.
- O_2 from blood diffuses across the membrane into the electrolyte filling solution and is reduced at the cathode.
- Circuit is completed at the anode, where silver is oxidized.
  - → Magnitude of the resulting current indicates the partial pressure of oxygen.
- Cathode reaction:

$$O_2 + 2H_2O + 4e^- \longrightarrow 4OH^-$$

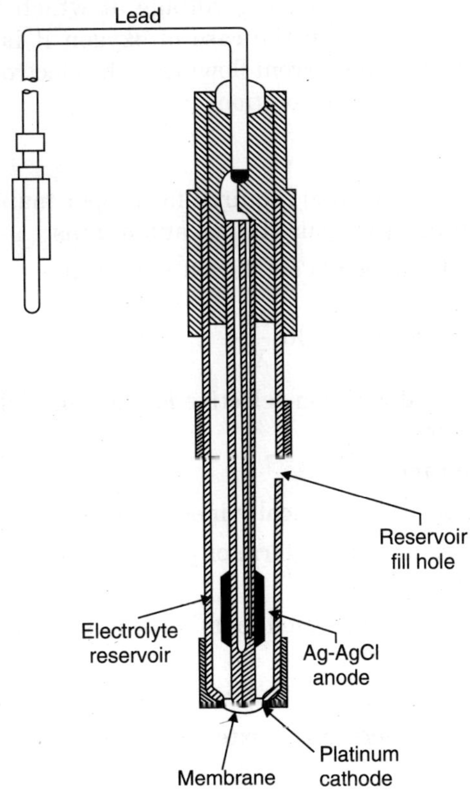

Fig. 3.9 Construction of pO_2 electrode (Courtesy: Corning Scientific Instruments, USA)

Anode reaction

$$4Ag \longrightarrow 4Ag^+ + 4e^-$$

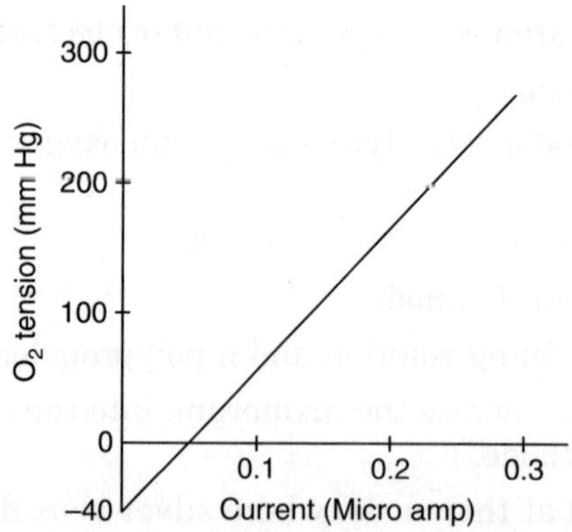

Fig. 3.10 Calibration curve of pO_2 electrode after Mc Conn and Robinson, 1963

Note. Do not pass through origin, but intersected the O_2 tension axis at a negative value.

- There is a characteristic polarizing voltage at which any element in solution is predominantly reduced and in the case of oxygen it is 0.6–0.9 V. In this voltage range, it is observed that the current flowing in the electrochemical cell is proportional to the O_2 concentration in the solution.

Ion Specific Electrodes

- There are several ions which is required for proper maintenance of health and also for proper functioning of various tissues and organs.
- Ion specific electrodes depends on Nernst potential

$$E_m = \frac{RT}{nf} \ln \frac{X_1^{\pm}}{X_0^{\pm}}$$

- Electrodes are provided with ion-selective membranes, allowing only the specific ion to pass through them.
- Ion selective membranes

  Have active agent in a chemically inert binder

  Active substances are in the form of paraffin, collodion, PVC, polyethylene (PE) and silicone rubber (silastic).
- Ion selective electrode is immersed in an outside solution (S_0) containing ion species $X_0^{\pm}$, which is to be measured

↓

Into this solution is immersed the external reference electrode

↓

The voltage E_{12} appearing across terminals 1 and 2 consists of 3 parts

1. Internal reference electrode in contact with the internal solution (S_i)—Half cell potential ($E_1 - S_i$)
2. Nernst potential (E_m) developed across the membrane
3. Half-cell potential – $E_2 - S_0$ → external reference electrode in contact with the outside solution (S_0)

$$E_{12} = E_1 - S_i + E_m + E_2 - S_0.$$

Ion selective membranes and ions they identify

Ion	Construction
K^+	Valinomycin/L (diphenyl ether) or PVC
Ca^{++}	Calcium di-n-decylphosphate/L (di (n-octyl-phenyl) PO_4) or PVC.
Zn^{++}	ZnSe + Ag_2S / disk
Mn^{++}	MnSe + Ag_2S / disk
Cd^{++}	CdS + Ag_2S / disk

GLUCOSE ELECTRODES

- Glucose is the essential metabolic input for many cells. The quantitative measurement of blood glucose is therefore very important and typically involves withdrawing blood samples, separating the plasma from the cells

  ↓

  adding a reagent to the plasma

  ↓

  processing the resultant solution to permit the calorimetric measurement–identifying concentration of the glucose.

  → This procedure is time-consuming and hence the need to create glucose sensors that provide an electrical signal proportional to glucose concentration.

Principle

When glucose is metabolised, Oxygen is consumed and gluconic acid and hydrogen peroxide are produced.

↓

In the presence of the enzyme glucose oxidase, the reaction proceeds rapidly.

$$C_6H_{12}O_6 + O_2 \longrightarrow H_2O_2 + \text{gluconic acid.}$$

It is therefore obvious that the glucose concentration can be measured by determining the oxygen consumed, the heat liberated, and the H_2O_2 or gluconic acid produced.

Enzyme Electrode

In these electrodes developed by Clark and Sachs (1968), the electrode consists of pO_2 electrode coated with enzyme glucose oxidase and covered with a cellophane membrane that is permeable to glucose and prevents oxidase from being poisoned by substances in the fluid being measured.

The decrease in pO_2 is a measure of glucose concentration.

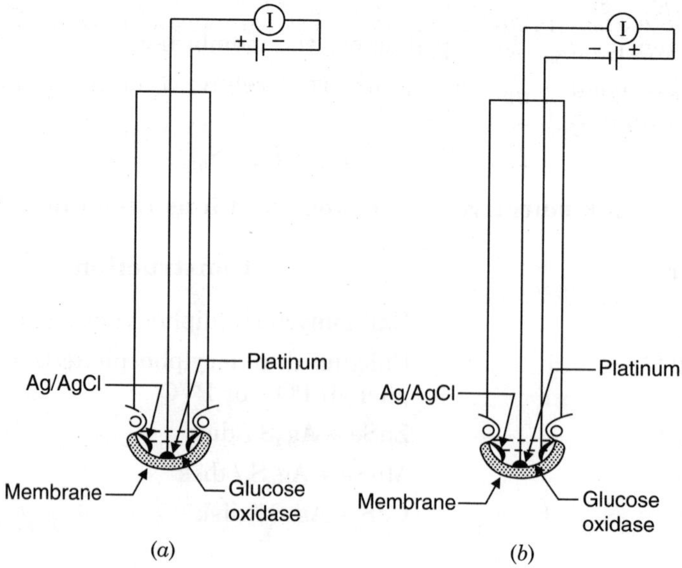

Fig.3.11 Enzyme electrodes for glucose. (*a*) The oxygen cathode, a Clark pO_2 electrode coated with immobilized glucose oxidase covered by a glucose-and oxygen-permeable membrane. (*b*) The peroxide electrode. In both cases the current *I* is proportional to the glucose concentration. [Redrawn from L.C. Clark, in *Theory, Design and Biomedical Applications of Solid-State Chemical Sensors*. P.W. Cheung and D.G. Fleming (eds.). CRC Press, Boca Raton, Florida, 1978, 296 pp.]

RECENT DEVELOPMENTS IN BIO MEMS

- The field of Bio MEMS has developed to a great extents in the recent times.
- Bio MEMS is being applied different areas for better, diagnosis and treatment of diseases, tissue repair etc.
- Some of the recent developments are

1. Single molecule manipulation and development

National Institute of Standards and Technology (NIST) has launched a multi-disciplinary project to develop new ways to measure the structure, function and behaviour of single biomolecules.

- A nanobiotechnology platform is under design to integrate nanopores, nanofluidic pathways, nanoparticles and other devices for isolating, manipulating and characterizing single biomolecules.
- The single molecules thus isolated are observed and analyzed, with wide variety of tools like 'Flourescence energy transfer – FRET, Surface Enhanced Raman Spectroscopy (SERS), ion transport etc.

2. Microfluidic devices

- These devices are more and more established in laboratory for biomedical research and will soon penetrate the diagnostic market for lab automation applications.

- This development is because of the capability of microfluidic devices to perform complete assay protocols with very low sample and reagent consumption, also at the same time there is minimum external interference.

- These microplates or microscopic slide formats can be designed to do steps like flow propulsion, dosing, mixing or washing required for carrying out assays.
- Thus these devices allow for simple handling of sample, incubation and detection.

3. Electronic taste chip

- Recently, this chip was developed by 'University of Texas' at Austin.
- This chip mimics the chemical features of the human taste bud.
- The chip has the capacity to analyze rapidly the chemical and biochemical content of complex fluids like Blood, (especially human) environmental sample and bioaerosol specimens.
- Technology is extremely versatile and can be used to measure electrolytes, protein antigens, antibodies, whole bacteria and DNA/RNA.
- These chips exhibit impressive analytical capabilities and diagnostic capabilities that can be compared to pH meter for acidity and ELISA for proteins.

  Yet they are compact in design and low cost, allowing for their use in numerous military and civilian applications.
- Another advantage is that since chambers are miniaturized and etched on chip – allows for multiple tests on a single chip.
- Testing requires a single drop of fluid and disposable cartridges.
- Most recently microbead arrays have been designed for detection of
  - → chemical weapon precursors
  - → Degradation products
  - → Identification of bacterial spores from the bacillus family.

4. Automated devices for sample preparation using microfabricated devices

- These microfabricated devices are being developed as an alternative platform for capillary electrophoretic applications.
- Designed for working with nano-to-picoliter volumes–which allows for high speed separations.
- Holds promise, especially for DNA sequencing.

5. Microstimulation system for Parkinson's disease

- Department of Electrical Engineering, University of California, Los Angeles
- Parkinson's disease is a severely debilitating and degenerative neurological disease.
- Though drug treatments are effective, prolonged use minimize the effects and leads to side effects – Dyskinesia (involuntary movements)
- Deep brain stimulation (DBS) can reverse the lesions deep within the brain
- This lab has designed, fabricated and tested a novel micromachine probe and stimulation systems in rats.

6. Bioartificial kidneys

- Due to various reasons, kidneys fail to perform their designated functions. The prevalent treatment at present is in center hemodialysis.
- However hemodialysis is more costly and offer inferior survival when compared to renal transplant.
- Division of Nephrology, University of Michigan Hospitals have developed a bioartificial kidney.
- It has completed phase I clinical trials in critically ill patients.
- The division is trying to engineer a successful miniaturized implantable kidney.

Musculoskeletal repairs

- If there is a tissue damage or death, there is no growth factor nor bone graft matrix available currently is capable of forming tissue unless it is placed in a site that contains a sufficient number of stem cells.
- Microfabrication methods allow design of unique microtextured surface with defined morphology, which can be planted with required number of stem cells/ progenitor cells.
- Bio MEMS allows for fabrications of such 3-dimensional matrics to get the desired results, Orthopaedic and Biomedical Engineering Cleveland Clinic found.

8. Implantable Bio MEMS for drug delivery

- Mircrochips/Bio MEMS containing and array of sealed, drug filler reservoirs have been developed and can be implanted in the body.
- Release of drug from the microchip's reservoirs can be controlled by pre-programmed mircroprocessors, wireless telemetry or biosensors.
- Successfully demonstrated both *in vitro* and *in vivo* by MicroCHIPS Inc., USA.

4

Valuing Nanobiotechnology

- Nanobiotechnology is a young and rapidly evolving field of research.
- It is a field of research at the cross roads of biotechnology and nanoscience.
- At present, nanobiotechnology is a field concerned with the utilization of biological systems like cells, cellular components, nucleic acids, protein to fabricate functional nanostructured and mesoscopic architectures comprised of organic and inorganic materials.
- Nanobiotechnology is also concerned with refinement and application of instruments to basic and applied studies of fundamental biological processes.
- Nanobiotechnology combines the contribution coming from analytical, bioorganic, and bioinorganic chemistry, physics, molecular and cell biology and material science.
- The focus of nanobiotechnology is on

  (*i*) **Biocompatible inorganic devices** for medical implants, microfluid systems—for lab-on-a chip applications for investigation.

  (*ii*) **Protein based nanostructures**—Crystalline bacterial cell-surface layers (S-layers) have been extensively studied and resulted in development of S-layer ultrafiltration membranes (SUMs)—for attachment of enzymes, antibodies, antigens, protein A, biotin, avidin *etc.*, in the development of solid-phase immunoassays and biosensors.

  Engineered nanopores—Applications in many areas like separation science, sensing and drug delivery.

  Polymer nanocontainers—Methods to develop hollow polymer particles. Have a high potential applications like confined reaction vessels, protective shells for enzymes, traps for PCR products, drug delivery *etc.*

  (*iii*) **DNA based nanostructures**

  - DNA protein conjugates—applied to the self-assembly of high affinity reagents for immunoassays, nanoscale biosensor elements, microstructured biochips *etc.*
  - DNA templated electronics.
  - DNA based metallic wires—To combine capabilities of biopolymers with optical, electronic and quantum properties to develop new generation devices at nanometer scale.

  (*iv*) **Nanoanalytics**

  - This field expected to produce major advances in medical diagnostics therapeutics, molecular biology and bioengineering.

- Nanoparticles are covalently linked to biological molecules such as peptides, proteins and nucleic acids—suited well as contrast agents for imaging, nanoscale carriers for drug delivery, nanostructured coating and scaffolds for medical implants and tissue engineering.

In short, the leading segments of biotechnology can be grouped as follows:

(*a*) Drug delivery
- Bioavailability.
- Sustained release.
- Targeted delivery.
- Targeted release.

(*b*) Imaging agents
- Magnetic Resonance Imaging (MRI).
- Near-Infrared Fluorescence Imaging (NIRF) Combined $\bar{c}$ MRI.
- Drug discovery.
- Rapid *ex-vivo* diagnostic.

(*c*) Biosensors
- Basic research.
- Implantable diagnostic devices.
- Internal diagnostics.
- Intracellular diagnostics.
- Massive funds promised for nanobiotechnology.
- Since the space race of the 1960s, only few areas have attracted media attention as much as nanotechnology.
- This has resulted in attracting large sums of funding from reactionary public and venture capitalists.
- Several agencies in USA expect nanotechnology market to be $ 1 trillion in 2015.
- Nanobiotechnology applications are predominantly in the research and development.
- Over $ 3 billion in public funding, available worldwide for nanotechnology research in 2003, with 8% dedicated to nanobiotechnology.
- Public and private investment are sponsoring research projects in academia and facilitating for establishment of spin-off companies exclusively focused on nanobiotechnology.
- Estimates of the total cost of biotechnology products represent $ 50 billion per annum, with nano aspects contributing to 1%.
- Rejuvention of human tissues $ 500 billion dollar—projected business *viz.*, Atherosclerosis, ostiosclerosis, liver cirrohsis, skin burns *etc.*
- Nanobiotechnology can play a big role in it.
- In US survey of manufacturing industry units showed ↑ significance of nanotechnology. In 2000, only handful of companies had nanotech programs. In 2005, 18% of surveyed industries were marketing nanotech incorporated products and about 80% expected to commercialize nanoproducts by 2010.

DRUG DELIVERY AND BIOSENSORS REVENUES

Drug Delivery Revenues

- There is an ever increasing market for advanced drug delivery system.
- Different advanced drug delivery systems/technologies.

(*a*) Polymers and Biomaterials

- Non-absorbable polymers
- Polyamine analogs
- Magnapore polymers
- Natural biopolymers
- Intelligent polymers

(*b*) Oral Drug delivery devices

- Extended—release technology
- Delayed—release technology
- Immediate—dissolution technology
- Other progress in oral delivery

(*c*) Pulmonary Drug delivery

- Metered dose inhalers
- Dry powder inhalers
- Nebulizers

(*d*) Drug-Eluting Stent technology, percutaneous drug delivery technology

- Intranasal delivery
- Iontophoresis
- Phonophoresis
- Electroporation

(*e*) Others

- Ophthalmic drug delivery
- Colloidal drug delivery technologies
- Site/organ specific drug delivery
- Needle-free drug delivery technology. Microscission—uses a stream of gas to bombard small areas of the skin with tiny 25 μm crystals.
- The global market for advanced drug delivery systems was more than € 37.9 billion in 2000 and € 75 billion in 2005.
- € 19.8 B, Controlled release
- € 0.8 B, Needle drug injection
- € 5.4 B, injectable/implantable polymer systems
- € 9.6 B, Transdermal
- € 12.0 B, Transnasal
- € 17.B, Pulmonary
- € 4.9 B, Transmucosal

- € 0.9 B, Rectal
- € 2.5 B, Lipo posomal drug delivery
- € 3.8 B, Gene therapy
- € 1.9 B, Miscellaneous
- Developments within this market—continuing at a rapid pace, especially in the area of alternatives to injected macromolecules.
- In USA, advanced drug delivery market is projected to reach $ 91 billion by 2009. The US drug delivery system is prospering and is transforming ordinary drugs into better drugs optimized for a broad range of applications.

  Reasons why the drug delivery market is rapidly expanding.
- At present, there are 30 main drug delivery products on the market.
- Total annual income for these drugs is approximately $ 33 billion, with an expected annual growth of 15%.
- Two major factors are responsible for this
  (*i*) Advances in diagnostic technology are outpacing advances in new therapeutic agents. Highly detailed information about a patient is becoming available, therefore promoting more specific use of pharmaceuticals.
  (*ii*) Advances of new drug formulation is expensive and slow.

- Estimates for the cost of developing a new chemical entity (NCE) from start to finish is about $ 1 billion.
- One product has been developed and is currently in Phase III clinical trials, called 'Exubera'—inhaled version of insulin by Pfizer and Nektar therapeutics. They expect annual sales of Exubera could be as much as $ 3 billion.

Biosensors Revenues

- First biosensor was developed in about 30 years ago, since there has been tremendous development, especially in the last decade the commercialization possibility has been realized.
- The importance of biosensor technology lies in its ability to simply and quickly measure biological and chemical substances.
- Expense of bringing biosensor—based product to market is tremendous, however considerable research is now focused on developing new products.

  In 1998, medical sales of biosensors → $ 692 million

  2003, medical sales of biosensors → $ 1.5 billion

  In 1998, total market for biosensors → $ 768 million

  2003, total market for biosensors → $ 1.7 billion.
- New analysis, projects world biosensors market $ 4.38 billion by 2011.
- Some focused applications of biosensors

1. Detection of DNA sequence on silicon chips—ideal biosensors should be fast, easy to use, specific and inexpensive.

A company called Thermo Biostar—a clinical diagnostics company, USA—developed an extremely sensitive method for rapidly visual detection of specific DNA sequences using silicon chips.

2. Instrument free target detection: Formation of a nucleic acid hybrid is transduced into a molecular thin film—can be visually detected in white light. Used to identify protein and carbohydrate targets in clinical samples associated with bacterial and viral pathogens.
3. Miniaturization: Miniaturized biosensors can sense any irregularity and designed to give loud and clear warning. Manufacturers have employed integrated biosensor and integrated chip technology to achieve this.
4. Development of Bacteria magnets for portable food and waters. Very useful in detecting adulterated/contaminated food and water.

BUCKYBALLS AND BUCKYTUBES

- Our understanding of C60 is from the observation of Harold W. Kroto. He discovered, naturally occurring unknown carbon in interstellar dust.
- Now they are believed to be throughout the galaxy and possibly surrounding "Red Giants" and "Black holes".
- Stephan Hanking's theories state that there is a form of black body radiation, which black holes produce. It appears that this is also characteristic of buckyballs. These fullerenes also can absorb and produce black body radiation.

Origin of Life

- Scientists have found gas filled buckyballs.
- These buckyballs were embedded in 4.6 billion-year old meteorites.
- If the theory of that atmospheric gases and organic compounds arrived on Earth during asteroid and comet strikes, then this can lead to further speculations that we are not alone in the universe, as these gases are required for development of life.

Buckyballs

- Were first noted in the stars, later identified as common soot.
- C60—produced mechanically in 1990. This was possible, when scientists were trying to create Giant star conditions.
- In nature carbon exists in many forms, most of us are aware about graphite and diamond, fullerenes also exist in nature.
- C60—also known as "Buckyball", other relatively common ones are C70, C76 and C84.
- Fullerene cages are about a billionth of a meter.
- C60, as the name suggests made of 60 carbon atoms, all of which are linked in a pentagon and hexagonal pattern—it's shape is identical to that of a soccer ball.

Buckytubes

In 1991, Sumio Iijima discovered buckytubes while working on buckyball.

- They are super fine filaments made of hexagonal "Chicken wire" carbon mesh.
- Both buckyballs and buckytubes, made of carbon, the buckyballs are made or look like balls, the tube rolls up and incorporates only 5-sided carbon at the ends of the tube.

- One can introduce 5-sided carbon at any part of tube and this considerably changes the tubes abilities and properties.
- They are very strong $\bar{c}$ high tensile strength. The strength is equal to a diamond. They can also conduct electricity.

Electrical conductivity	—	Found to be best conductor of electricity on a nanoscale level, as good as copper.
Thermal conductivity	—	As good as diamonds.
Mechanical	—	Strongest and toughest fiber that has been found yet.
Molecular perfection	—	They have a high degree of uniformity and freedom from defects.
Self-assembly	—	Because of strong forces, they can automatically group together, under certain conditions.

- Last year (2004) only about 3 kgs of nanotubes were produced, tagged $\bar{c}$ a price tag of Rs. $ 300 per gram, which is 30 times the price of gold. Analysts are hopeful that soon, the prices may fall down to $ 50–60 per gram.
- Companies such as IBM, Xerox, Mitsubishi and DuPont have patents on buckyballs, and continuing their experiments on new concepts.

Electrical Charges

- This is another method for moving particles through micro-channels is 'Electrokinetics'—
  1. Electrophoresis—Moving particles three nanochannel by placing an external voltage difference across the entire channel.
  2. Electroosmosis—Similar to electrophoresis, however here the sides of the channel wall is also charged. Charged particles in the liquid interact and slip past the charges on the channel wall.

Fabrication of Microfluidic Channels

1. A master is created on a silicon wafer, using traditional processor-fabrication methods. To get a raised pattern that will act as a mold for our polymer, we use a *negative* photoresist: We etch away the surrounding material till what's left is a raised pattern that outlines the desired channels on the silicon wafer.
2. We then cure a polymer onto our silicon wafer mold, which now has a channel imprint with nanometer-scale features. *Curing* involves adding heat to make the liquid polymer harden a bit—ultimately it feels and looks like Jelly—this is Soft Lithography. The polymer is polydimethysiloxane—PDM
3. Peel the cured layer of polymer from master.
4. Oxidize layer of polymer, thus sealing it to another layer of either PDMS, Silicon/ glass.
5. Add holes—an injection reservoir, to the polymer, allowing us to inject liquids.

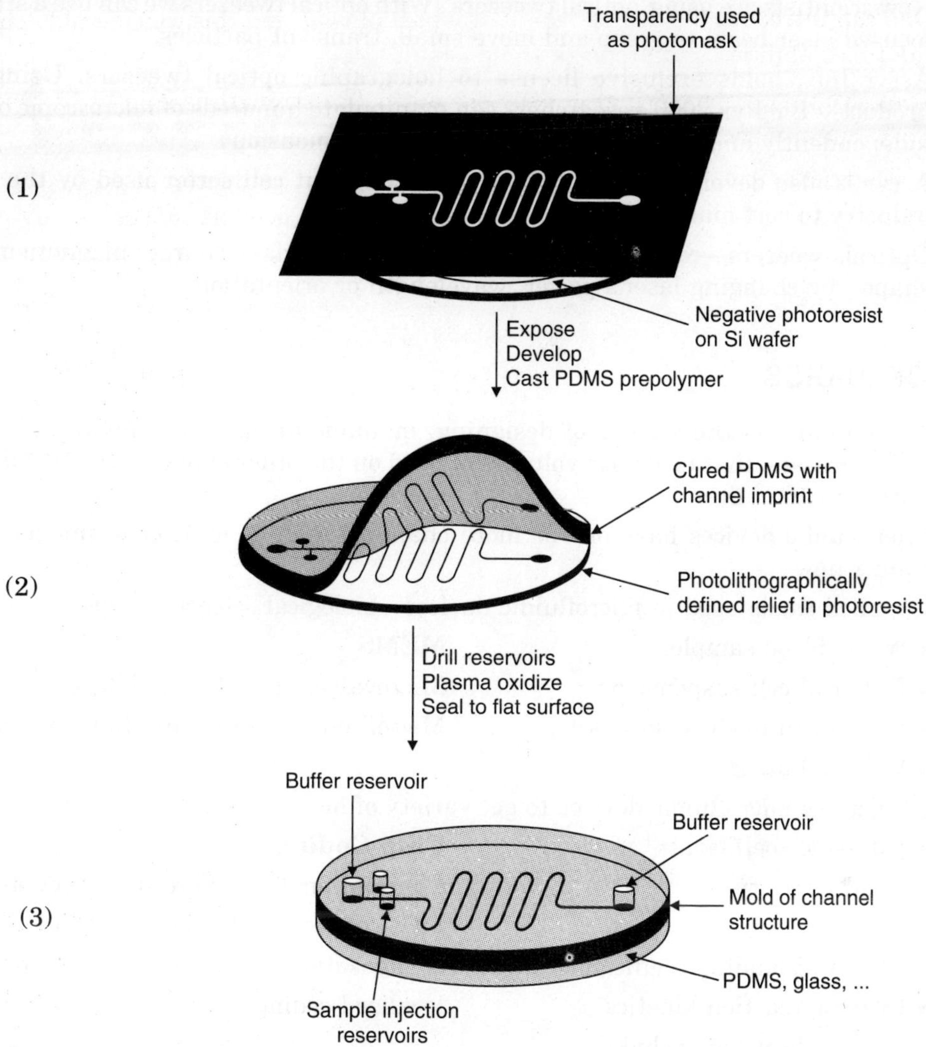

Fig. 4.1

Surface Tension

- Imagine a water droplet on a wax material.

↓

This water droplet is bulk inside, with a surface shell outside.

↓

Bulk inside has water molecules interacting with each other, the surface molecules have to spend more energy in doing that, larger the surface area greater is the energy cost.

↓

Surface tension is the pull of a liquid into its most compact form, to minimize the amount of energy used.

One can control the surface with light.

- Scientists at Tokyo institute of technology, have used UV laser to change the chemical structure of a droplet's surface, allowing the spread out the droplet, by using blue laser they were successful in bringing droplet tighten together.

- Now scientists are using 'optical tweezers'. With optical tweezers we can use a strongly focused laser beam to grasp and move small, translent particles.
- Arryr Inc., holds exclusive license to holographic optical tweezers. Using this product—'Bio Ryx 2000' researchers can manipulate hundreds of microscopic objects independently and simultaneously, in all three dimensions.

  Arryr is also developing Mat Ryx—high-throughput cell sorter used by the cattle industry to sort male and female reproductive cells.
- Optical tweezers—can help to sort cells by size, surface charge, magnetism, and shape—by changing laser's power, wavelength or orientation.

MICROFLUIDICS

- Microfluidics is the science of designing, manufacturing and formulating devices and processes, that deal with volumes of fluid on the order of nanolitres 10^{-9} liter, or picoliters 10^{-12} liter.
- Microfluidic devices have one or more channels with at least one dimension less than 1 nm.
- Common fluids used in microfluidic devices—biological sciences include
  - Whole blood samples
  - Bacterial cell suspensions
  - Protein/Antibody solutions
  - Various buffers

MEMS

Microvalves, Flexible capillaries, Microfluidic pumps, Thin film coating.

- One can use microfluidic devices to get variety of measurements like
  - Diffusion coefficients
  - Fluid viscosity
  - p^{17}
  - Chemical binding coefficients
  - Enzyme reaction kinetics

Chip Coding

Thermal analysis, Forced convection, Free convection, Evaporation and condensation, Inkjet printing based localized coding.

- Other applications include
  - Capillary electrophoresis
  - Iso electric focusing
  - Immunoassays
  - Flow cytometry
  - Protein analysis for/via mass spectrometry
  - PCR amplification
  - DNA analysis
  - Cell manipulation
  - Cell separation, cell patterning thus playing a very important role in clinical diagnostics.
- When one works at the micro-scale level, particles in the fluid will be almost the same size of device itself—this fact changes the basic physics of fluids dramatically.

- At micro-scale level, surface-to-volume ratio is much greater. Therefore, properties such as viscosity, surface tension and electrical charges change drastically.

Viscosity

- Is the measure of resistance of a fluid *i.e.,* in simple thickness of the liquid.
- Fluids at micro and nano level are highly viscous—they move more like honey than water.
- This results in a property known as "Laminar flow", preventing fluids from mixing readily. With this, two liquids come together they stay parallel to each other as they go down the channel, flowing smoothly and without turbulence.

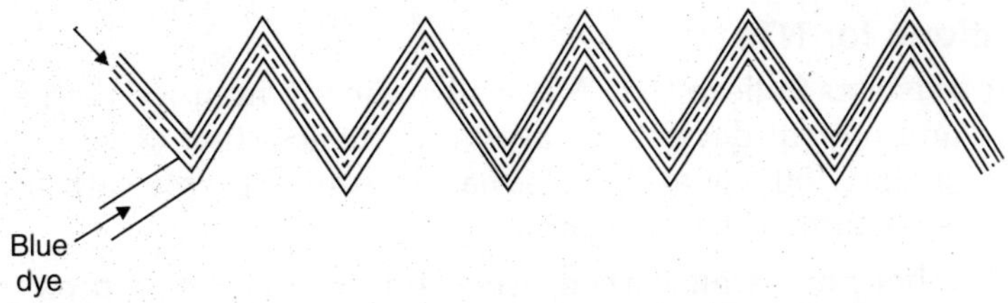

Fig. 4.2

Agitent technologies and Caliper technologies have collaborated to exploit laminar flow to help speed up drug discovery.

↓

The technique used cells with fluorescent dye and allowed to interact with different chemicals. The dye glows when the new drug reacts with proteins on the cell's surface—a sign that drug can enter the cell.

NANOTUBES vs ENVIRONMENT

Nanotubes is one of the structures most commonly identified with nanotechnology in the popular press under the name of "carbon nanotube" or "buckytube".

Nanotubes are currently subject of huge enthusiasm because of their amazing potential for our technological future. But amidst all this excitement we tend to forget the significant impact these materials may have on our environment. These risks should not be taken lightly. If the studies and uses of these materials are not heeded and the potential downsides not carefully monitored or appropriately handled, we may be faced with the development of previously unknown cancers or diseases.

What are Nanotubes (NTs) ?

NTs are large molecules of pure carbon that are long and thin and shaped like tubes, about 1-3 nanometers (1 nm = 1 billionth of a meter) in diameter, and hundreds to thousands of nanometers long. As individual molecules, nanotubes are 100 times stronger-than-steel and one-sixth its weight. Some carbon nanotubes can be extremely efficient conductors of electricity and heat; depending on their configuration, some act as semiconductors. They come in several varieties: long, short, single-walled or multi-walled, open or closed with different types of spiral structures. Each variety is used for specific requirements and applications.

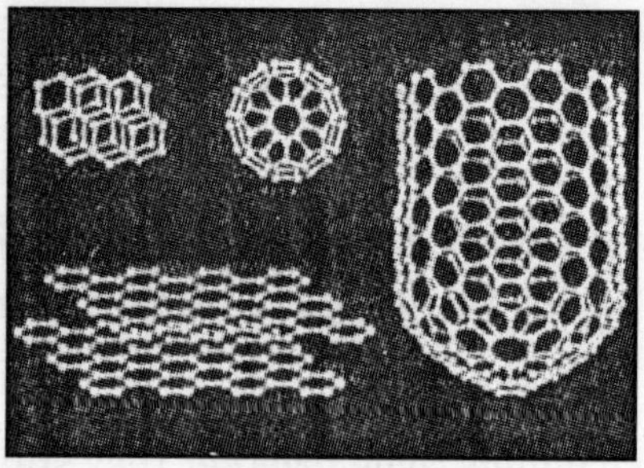

Fig. 4.3 NT model

Industry Applications for NTs

Industry enthusiasts believe that NTs will radically improve the performance of tiny sensors, electronic and optical devices, catalysts, batteries, fuel cells, solar cells and drug delivery vehicles. Currently 50% of all lithium batteries incorporate carbon nanofibers (wires spun from NTs), which double their energy capacity.

Some predict that nano-scale carbon transistors will replace silicon transistors within the next decade. NTs are already used in such things as tennis rackets to make them stronger and lighter. Fortifying a bulletproof vest with a small quantity of NTs could double its ability to absorb the energy of a bullet. One company is developing NTs to make plastics fire retardant. NTs are capable of storing up to 65 per cent of their weight in hydrogen—a capacity that could someday make hydrogen fuel cells a cheap and efficient alternative to fossil fuels. Scientists at Rice University are developing a new type of wire made of NTs that conducts electricity much better than copper, and could transform the electrical power grid.

Some examples where such NTs are used are as follows:

- NASA researchers have reported a new method for producing integrated circuits using NTs instead of copper for interconnects.
- NTs are being used for drug delivery and other nanofluidic applications.
- Rensselaer Researchers use NTs to create efficient filters.
- International Sematech and Xidex Corp. have developed a scanning probe microscope (SPM) technology, based on NT technology.

In the world of Nanotechnology, NTs are the stars because of their tremendous properties and great potential for tomorrow's market. Even if they are still expensive, their prices will most likely decrease dramatically in future.

There is no doubt about it, the production of NTs will increase over the next few years and we will be seeing this new material everywhere: in our cars, architecture, planes, steel, plastics, *etc.*

Can NTs be used safely with our natural environment?

Anything you work with today can have an effect our environment or personal health in some way. This is more pronounced when you are working at the micro-scale or nano-scale level. Such was the case with the asbestos.

Due to its strength and resistance to heat and chemicals, Asbestos was often used in older building materials (floor tiles, ceiling tiles, insulation on pipes and ducts, acoustical and decorative coatings, and roofing materials). When left intact and undisturbed, these materials do not pose a health risk to building occupants. There is a potential for exposure only when the material becomes damaged to the extent that asbestos fibers become airborne and are inhaled. Asbestos is more likely to release fibers when it is friable. The term friable means the material can be easily crumbled. If powdered or friable forms of asbestos are disturbed and become airborne, an inhalation hazard may result.

NTs are at the cross roads with bright promises and expanding horizons and downside factors are coming to the light.

Can bath NTs and Environment be handled?

Despite the huge amount of interest and investment in NTs, the toxicological impacts of these and other engineered nanoparticles are still unknown. A handful of toxicological studies reported thus far indicate that there is reason for concern.

I refer to an article published, on November 18th, 2003 by ISIS (Institute of Science in Society) entitled "Nanotubes Highly Toxic" where scientist discovered that NTs could be harmful for the health if not using with cautions.

"Unprocessed nanotubes are very light, and could become airborne and potentially reach the lungs. Researchers in the Space and Life Sciences of NASA Johnson Space Center, Wyle Laboratories, and the Department of Pathology and Laboratory Medicine, University of Texas Medical School, in Houston, Texas, USA, investigated the toxicity of carbon nanotubes to the lungs, by introducing them into the trachea of mice under anaesthesia.

The results are alarming. Five of the mice treated with high dose of one kind of nanotubes died within 7 days. All nanotube products induced epitheliod granulomas-tumour-like nodules of bloated white blood cells in the lining of the lungs—and in some cases inflammation of the lungs at 7 days. These persisted and became more pronounced in animals that were sacrificed at 90 days. The lungs of some animals also showed inflammation around the bronchi, and extensive necrosis (tissue death).

Carbon nanotubes, the researchers conclude, are "much more toxic than carbon black and can be more toxic than quartz, which is considered a serious occupational health hazard in chronic inhalation exposures."

This other article was published March 10th, 2005 by "The National Institute for Occupational Safety and Health" and describes the same potential danger that can result if one was exposed to NTs.

"If workers were exposed to NTs at the level of the PEL for synthetic graphite it is likely there would be some lung lesions," said Anna Shvedova, who works in NIOSH's pathology and physiology research branch.

NIOSH's scientists are finding that carbon nanotubes cause a variety of lung and cardiovascular effects, according to three presentations the scientists gave at SOT. For example, granulomas—inflamed cells that clump together-develop rapidly in the lungs of laboratory mice exposed to carbon nanotubes, Shvedova said...

James, from NASA, repeated a recommendation made earlier in the meeting that until more information is available about the health risks nanoparticles pose, work sites should be very careful about protecting their workers."

As a scientist, I believe that NTs can help make our world a safer and better place if we use them wisely.

Nature has own way for creating balance. Whenever, there is a great gain in something, there typically are counter factors that come into play. (This debate could be another subject that can be addressed in another article).

NANOCRYSTALS

Nanocrystal Imaging Corporation

Why Digital X-ray Imaging ?

X-ray imaging (which includes CT-SCAN and fluoroscopy) accounts for nearly 80% of all medical imaging. The advantages of capturing and storing X-ray images digitally rather than on film are overwhelming. Digital X-ray imaging has many advantages:

(*a*) The images can be stored in a very compact manner without risk of loss or deterioration in quality.

(*b*) The images can be easily enlarged and "suspicious" areas of an image can be more easily studied.

(*c*) The images can be easily retrieved, and can be transmitted electronically to an expert or to another medical practitioner in the same hospital or to an insurance company for reimbursement purposes.

(*d*) Over a period of time, pattern recognition software can be developed to aid easy diagnosis.

(*e*) The expense of X-ray film and the chemicals used for development as well as the environmental problems associated with the disposal of film can be entirely avoided.

Problems Associated with Digital X-ray Imaging

X-rays are a form of energy which do not fall within the visible light spectrum, and must first be converted to light in order to be recorded in either digital or analog form. X-rays used for medical imaging are scattered in all directions at two distinct points of impact: first, by the anatomical features being imaged, and then by a screen (the scintillator) which converts the X-rays emerging from the body into light. As at present, there is no known method to minimize the scattering of X-rays by the anatomic features being imaged. However, X-ray scattering at the second point of impact (*i.e.*, at the scintillator) can be minimized by collimating the emerging X-rays. Another problem in X-ray imaging has been the efficient absorption and conversion of X-rays by the scintillator before the X-rays.

Fullerenes, Carriers and Dendrimers

- All conventional materials like metals, semiconductors, glass, ceramic or polymers can be obtained in nanoscale dimension.
- Spectrum of nanomaterials ranges from inorganic/organic, crystalline or amorphous particles.
- Fullerenes and other supramolecular structures like dendrimers, micelles or liposomes belong to the field of nanomaterials.

Dimension

Single < 100 nm	Films, coating, multilayer
Two < 100 nm	Tubes, fibres, wires, platelets
Three < 100 nm	Particles, quantum dots, hallow spheres *etc.*

Nanowires and Carbon Nanotubes (Nanorods)

- Linear nanostructures like nanowires, nanotubes or nanorods can be produced from various materials *e.g.*, metals, semiconductors or carbon by several techniques.
- Carbon nanotubes can occur in variety of modifications—singlewalled, multiwalled, filled or surface modified.

- Can have several applications—data storing, wiring, and also as filler for nanocomposites.

Nanopores

- Materials with defined pore sizes of nanometer range are of special interest.
- They show outstanding properties like thermal insulation, controllable material separation and release.
- Broad range of potential applications include catalysis, thermal insulation electrode materials, environmental filters and membrane and for control release drug carries.

Dendrimers

- Dendrimer, from the Greek word (dendron) for **tree,** refers to a synthetic, three dimensional molecule with branching parts—basically resembles a foam ball with tree springs shooting out in every direction.
- Dendrimers are formed using a nano-scale, multistep fabrication process.
- Donald Tomalia, first synthesized and named dendrimers in 1979.
- A typical dendrimer is the size of a typical protein—3 nm across, held together with strong chemical bonds—also containing of a great number of empty spaces/pockets to carry a payload.
- These empty spaces make up a great amount of surface area, and can be tailored to different sizes allowing for wide variety of applications, including diagnostic and therapeutic applications.
- One such application—transport DNA into cells for gene therapy. This is an alternative to troublesome genetically modified viruses. Dendrimers, covertly bypass the immune defenses and slip DNA into target cells.
- By customizing and controlling dendrimer architecture, nanotechnologists are developing dendrimers for drug delivery.
- Given their ability to be controlled, they may be designed to swell and liberate their contents only when the appropriate trigger molecules are present.
- They may be used for diagnostics—they can be made to deposit a type of DNA/chemical that makes a diseased cell glow.
- Companies producing them claim that most dendrimers don't trigger the immune system, when injected or used topically and have low cytotoxicity.

Other Applications

- Dendrimers could also be used in coatings and materials, electronics and photonics.
- Many patents are assigned to use of dendrimers in chemical, petroleum, tyre, cosmetics and pharmaceutical companies.
- Commercial development of dendrimers has been slow, because of difficulty of scaling up production and their cost is very high.
- Diagnostic grade dendrimers may cost $ 1,650 per 100 mg.
- However, few patents on a new, one-step process to synthesize dendrimers can drive down the cost of productions.

Dendrimer Based Products

- A dendrimer based tool for detecting cardiac image is being developed by Dade Behring, one of the world's largest medical diagnostic firms.
- World's first dendrimer based drug, developed by Australian-based Starpharma is a topical gel for use as a "liquid condom" to reduce the risk of HIV infection in women.
- US Army Research Laboratory is developing a dendrimer-based Anthrax detection agent, dubbed "Alert Ticket".
- ExxonMobil owns a patent on a "Flowimprover" based on dendrimer technology—an additive that will enhance the flow of oil in cold temperatures.

OPTICAL/PHOTOELECTRIC TRANSDUCERS

- Play an important role in measuring physiological events in living subjects and are employed in two ways.
- First method—photosensor detects changes in the light intensity of a given wavelength like in colorimetry/spectrophotometry.
- Second method—detects changes in the intensity of light where wavelength is relatively unimportant like scintillation counter.
- There are 3 basic types of photoelectric transducers
  - Photoemissive tubes
  - Photovoltaic cells
  - Photoconductive cells

1. Photoemissive tubes

- Consists of an evacuated/gas filled bulb with two electrodes

  ↓

  One electrode is cathode, which is designed to release electrodes

  ↓

  The other is anode, consists of a thin rod/loop of wire.
- Electrons are released only when cathode is hit by a wavelength of light not longer than the threshold level.
- Emissive surfaces are usually cesium, antimony, silver and bismuth in combination.
- A relatively high voltage (10–200 V) must be applied between two electrodes. e^- released are attracted to the anode.
- In the vacuum type current produced is small. For higher current requirements, gas mixtures are often employed.
- Both vacuum and gas filled photoemissive detectors respond quickly to changes in light intensity—former 10^{-9} sec, latter 10^{-3} sec.

2. Photovoltaic cells

- Photovoltaic/photogalvanic cells—frequently encountered in biomedical studies.
- They do not require high voltage like photoemissive cells, but develop voltage high enough to be measured three galvanometer/other low impedence circuits.

- These cells usually contain a sandwich of selenium between iron or steel backing.

  ↓

  Above selenium—a thin transparent film of metal is present and these two are insultated from one another acts as barrier layer.

  ↓

  When barrier layer is illuminated, light quanta are absorbed and e^{-}s are released and potential difference appears across the barrier.

  ↓

Transparent film becomes (–ve) and selenium (+ve)

- CuO in contact with gold/platinum also employed in barrier layer cells.
- Selenium photocells cover visible spectrum 300–700 nm.
- With an inexpensive filter it allows us to obtain spectral curve similar to human eye therefore used in illumination meters, exposure meters, colorimeters *etc.*
- Response time of a typical voltaic cell is usually less than 5 m sec.
- P-N junction type semiconductors are employed to convert sunlight into electric power.
- Since junction photocells's response time is shorter, small in size, high sensitivity used in life sciences. They are used to detect the light produced by a light-emitting diode (LED).

3. Photoconductive cells

- Photoconductive/photoresistive cells consists of a thin film of a material *viz.*, selenium, germanium, silicon or a metal halide/sulfide.

  ↓

- When exposed to certain types of radiant energy, it exhibits the photoconductive phenomenon *i.e.,* decrease in resistance.
- Most photoconductive cells will exhibit a drop in resistance from many megohms in the dark to a few hundreds of ohms when highly illuminated.
- Many of the photoconductive cells show good sensitivity to visible light and a few to UV and X-ray spectra. A large number are sensitive to IR region therefore used in development of spectroscopy and self guiding IR seeking missiles.
- Typical response times for photoconductors operated at room temperature vary between 0.1 and 30 m sec.

Scintillation Counter

- It is an example of photomultiplier—photoemmissive transducers.
- 'Scintilla' is a flash of light.

  ↓

  A scintillation counter counts number of light flash—usually employed to monitor the decay of redioactive elements.

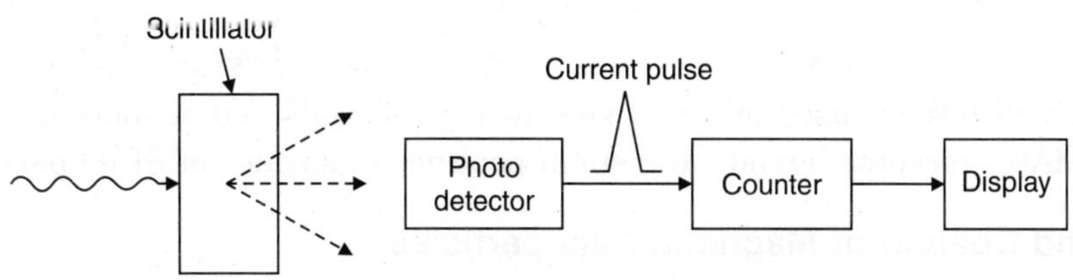

Fig. 4.4

- When an energetic particle *viz.,* alpha particle/photon strikes certain crystals (or some materials dissolved in liquid), a faint flash of light is produced.

  ↓

  Scintillation—detected by a photomultiplier tube—which delivers a short pulse of current to the counting circuit that accumulates and displays the counts.

N.I. and Z.SO. as crystals, benzene, naphthalene or anthracene also used as scintillalors.

TARGETING AND FUNCTIONALIZATION OF NANOPARTICLES

- One of the leading areas in use of nanoparticles—Medical imaging.
- 'Magnetic Nanoparticle Probe'—emerging class of novel contrast and tracking agents for medical imaging.
- SPIOs—Superparamagnetic iron oxide nanoparticles—allow researchers and clinicians to enhance tissue contrast by the relaxation rate of water.
- SPIOs—available in different sizes, coating thickness, surface chemistry and targeting ligands—can be tailored to target specific organs, cells or even molecular markers of different diseases '*in vivo*'.
- Iron oxide nanoparticles—most often are magnetite (Fe_2O_3/Fe_3O_4) and show unpaired spins—but these superparamagnetic crystals show orderliness of magnetic domain when external magnetic field is applied—results in magnetic moment, which is measured.
- SPIO particles—differ in core size, coating thickness, coating material—and these can dictate where SPIOs end up '*in vivo*'.

Example: Very large particles—300 nm to 3.5 μ—coated $\bar{c}$ insoluble layer—used to image GI tract.

Particles—60 nm to 150 nm—get into reticuloendothelial system (RES) upon injection—rapidly appearing in the liver and spleen.

Much smaller particles 10–40 nm—circulate in the blood for longer time—cross capillary walls—often taken up by lymph nodes and bone marrow.

- Differential cell uptake in cancerous tissues leads to contrast enhancement between normal and tumar tissues.
- SPIOs are used for **'cell tracking'.** Cells, usually stem cells—labelled by SPIO prior to '*in vivo*' linery. This allows specific cells to be tracked over a period of years to weeks to
  (*i*) determine cell migration
  (*ii*) efficiency of cell delivery and incorporation
  (*iii*) other parameters for assessing success of cell delivery techniques.

Recently SPIOs are used to target specific tissues, cells and biomolecules by attaching ligands *viz.,* MAbs, receptor ligands and small peptides to surface of SPIO particles.

Synthesis and Coating of Magnetic Nanoparticles

- There are various chemical methods adapted synthesizing magnetic nanoparticles for a wide range of applications.

- Methods

    (*i*) Precipitation based synthesis

    Coprecipitation | Reverse micelle synthesis

these have certain problems like low soluble in water, therefore coating method is adapted.

Coating not only makes them **water soluble,** also prevent them undergoing **'aggregation'.** In addition to this one conjugate $\bar{c}$ biomolecules for **targeting.**

- Most common method adapted for coating—coprecipitation in the presence of **'polysaccharide dextran',** coating further crosslinked chemically to enhance stability.

    ↓

    These SPIOs are more biocompatible, water soluble and can be used for wide range of preclinical and clinical imaging studies. Other coatings include:
    - Carbodextran coating
    - Starch based coating
    - Dendrimer based coating.

Post coating use variety of materials *viz*., (*a*) monolayer ligands, (*b*) polymers, (*c*) combinations of polymers and phospholipids + carbohydrates and (*d*) Silica coating.

Example: Particles $\bar{c}$ thick coatings composed of polymer → polystyrene-coated SPIO—oral magnetic particles (OMP) and siloxane—mainly used for GI track MRI.

PEG—phospholipid coating $\bar{c}$ bioconjugation of various moieties like proteins, oligonucleotides and delivery peptides—used for intracellular targeting of specific proteins and nucleic acids.

Biomedical Applications

Agent	Particle size	Target organs	FDA approval
AMI-121	300 nm	GI Lumen	Approved
OMP[50]	3.5 μ	GI Lumen	Approved
AMI-25	80–125 nm	Liver/spleen	Approved
SHU555A	62 nm	Liver, blood pool	Phase III
NC100150	20 nm	Blood pool	Under development

All medical applications—rely on contrast enhancement of anatomical areas to which SPIOs are naturally directed by physiological processes—To enhance more specificity—direct targeting by coating SPIOs $\bar{c}$ ligands/other biomolecules.

Cell Labelling and Tracking

- This application a very successful, after medical imaging.
- Ability to load enough nanosized particles $\bar{c}$ micromolar iron particles in cell culture methods using various methods *viz.,* cell permeable peptides and transfection agents one is able to label and track non-phagolytic cells by MRI.

- Especially used in, to study biodistribution of stem cells after implantation, monitoring their migration and trafficking and determine success of cell-transplantation therapy over 6 weeks.

NANOTUBES AND NANOCRYSTALS

Nanotubes Vs Environment

- Nanotubes—one structure—commonly identified $\bar{c}$ nanotechnology "Carbon nanotube" or "Bucky tube."
- Currently subject of huge enthusiasm because of their potential applications—But amidst all this—one needs to look the significant impact of these materials may have on our environment.
- If we do not carry out proper studies and research and carefully monitored or appropriately handled—the world may face $\bar{c}$ development of unknown cancers/diseases.
- Despite huge amount of interest and investment in NTs—toxicologial impacts of these and other engineered nanoparticles—still unknown.
- Few studies already made, gives an indication to be cautious.

  Article published, 18th Nov. 2003 by Institute of Science in Society—ISIS, entitled "Nanotubes Highly Toxic," where it has been reported NTs could be harmful for health if not used judiciously and cautiously.
- Unprocessed nanotubes are very light—could become airborne and potentially reach the lungs.

↓

Researchers from NASA Johnson Space Center, USA, Wyle laboratories and University of Texas Medical School, USA—investigated toxicity of C[1]—Nanotubes to the lungs introducing them to the trachea of Mice under anaesthesia.

↓

Results alarming → 5 of mice treated $\bar{c}$ high dose of nanotubes died in 7 days.

↓

All nanotubes induced epitheliod granulomas-tumor like nodules (bloated wBls in the lining of lungs) $\bar{c}$ inflammation

↓

Lungs of some animals showed inflammation around the bronchi and extensive necrosis *i.e.,* tissue death.

- Researchers concluded carbon NTs to be more toxic than **carbon black** and can be more toxic than **quartz.**
- 2nd Article. Published in March 10th, 2005 by "The National Institute for Occupational Safety and Health"—describes once again potential health hazard.

If workers are exposed to NTs at level of synthetic graphite—likely there would be some lung lesions.

Scientists have also concluded that G-NTs can cause a variety of lung and cardiovascular diseases.

Therefore, though NTs can help make our world safer and better place—we need to use them wisely.

Nanocrystals

- These are nanoparticles, showing crystalline structure and emit spectrum of colours depending on size and shape.
- Have promising application in digital X-ray imaging.

X-Ray Imaging

- Accounts of nearly 80% of all medical imaging.
- Storing images digitally over a film has many advantages like—can be easily retrieved and transmitted, can be stored in a very compact manner, allows for enlargement of suspicious areas.
- However certain problems are associated with digital X-ray imaging.
- X-rays do not fall under visible light therefore needs to be converted to light in order to be stored digitally.
- X-rays during imaging undergo scattering in all directions during
  (*a*) By the anatomical structures—till today there is no known method to minimize this scattering.
  (*b*) When they hit the scintillator—which converts X-rays into light.
- Subsequently converted light stored on a **CCD**—Charged Couple Device or **CMOS**—complementary Metal Oxide Semiconductor.
- NIC has developed novel methods to solve the problem and at the same time to get high resolution pictures/images.
- NIC has successfully created a *novel scintillator* based on a microchannel plate containing between 2-3 million channels/sq.inch.
- Each channel contains NIC's proprietary bond of highly efficient phosphors.
- A high resolution digital imaging is obtained by integrating this scintillator with CCD or CMOS.
- High resolution efficiency—particularly important in the mammography and dental X-ray imaging.

Other areas

Osteoporosis—Bone density mapping. Micro CT-scan can help in better imaging and to understand impact of drug.

Cardiac Imaging—High resolution and high contrast X-ray images would help—cardiologists to perform catheterization and placement of **stents**. In future one aims to observe **plaque** formation.

Oncology Imaging—High contrast and resolution—can help medical community to locate the boundaries of cancer growth for radiation therapy treatment.

With Nanotechnology it may be possible to develop an affordable desktop digital X-ray imaging systems—which depend on very low X-ray dosage, hence any average physician will be able to use it in his clinic.

Doped Nanocrystals

- These are coaxed nanocrystals that are made to accept foreign materials.
- These have an important role to play in the field of electronics.

- They can be doped $\bar{c}$ impurities like **phosphorus** that alter conductivity and other electronic properties in a controlled way.
- A team of scientists at the 'Naval Research Lab—NRL have developed a model to prepare manganese—doped cadmium selenide nanocrystals.
- Principle of kinetics is applied while creating such doped nanocrystals. Foreign material can be incorporated to the nanocrystal as it grows during synthesis.

Nanoscale Biosensors

Let us try to understand importance of sensors, by taking few examples:

- If food processing industry could tell whether food adequately exposed to heat or has begun to spoil.
- A tiny foolproof automatic indicator—that could constantly monitor home/industry for hazardous gas leaks, CO and other dangerous chemicals.
- A simple swab test—that can instantly tell us whether patient has strep infection, diabetes and other various genetic diseases.
- A simple wire that could be inserted in the ground to tell a gardener whether the soil is suitable for planting a particular species.

All these possibilities are potential applications of **Nanosensors.** By definition from **Oxford English Dictionary—**Sensor is a device giving a signal for the detection or measurement of a physical/[chemical] property to which it responds.

↓

But application of sensors to the detection of molecular structure is perhaps the most promising and important area of sensing activity.

Drug Discovery Using Nanocrystals

Nanoenabled drug discovery will dominate:

- Nanotechnology for molecular targeting and drug delivery is set to become a **Trillion dollar** industry.
- The scale of funds indicate—how seriously 'Nanotechnology' is being taken by countries.

  Example: In 2003—US spent $ 860 million, whereas $ 650 million by Europe, Japan about $ 800 million.

  National Cancer Institute—had a budget 2 $ 144 million in 2005.
- Nanotechnology is bound to make big impact in the field of drug research and development.

  This is believed to help pharma industry is the face of growing **regulatory** and **pricing pressures.**
- There is also increase in levels of MOs developing resistance to antibiotics and cancerous cells not responding to the drugs.

- One of the prominent technique that could make a prominent mark is **'AFM'**.

↓

The current, microscopic techniques—Unable to see how components of a cell react in biological process—in presence of a specific chemical/compound.

↓

AFM allows for attaching specific Abs to individual proteins acting as probe—when Ab reacts $\bar{c}$ target protein—creates variance in microscope's reading, showing protein's presence.

↓

This technique in future will allow for detailed understanding of the chemical dynamics involved in cell's reaction to a particular.

- Presently the technology not widely used as it is hindered by limited throughout and experimental burden of complicated and expensive instrumentation.

↓

- They are predicting by using submicrometer cantilever arrays may ↑ throughout of AFM based spectroscopy.

↓

Such AFM based studies can improve the techniques of imaging—contributing significantly to drug discovery.

- Near-Field Scanning Optical Microscope **(NSOM)**—allows for studying sample surfaces—optical properties—resulting in a study with better resolution that wavelength of light.

↓

They result in production of "evanescent waves" from the surface and are detected by probes.

- The focus of nanotechnology is thus to improve imaging techniques which ultimately helps in drug discovery.
- **Surface Plasmon Resonance (SPR)** is one such recent development in image development.

SPR is a phenomenon—occurs when light is reflected off thin metal films.

↓

During the process—small amount interacts $\bar{c}$ $\bar{c}$s in the film—reducing the light intensity.

↓

Materials sandwiching the film—due to their refractive index dictates the angle at which light reduction occurs—resulting in a shadow.

↓

Using SPR based approach—interaction of biomolecules can be detected, allowing biological system action to be studied at small scale (Nano).

↓

Successfully employed by 'Applied Biosystems'—its 8500 Affinity chip analyser uses SPR to measure binding of biomolecules to biostimulants.

Dip-Pen Nanolithography—DPN

- This is other art of writing and printing at nanoscopic level.
- Nano Ink—pioneered in DPN—provide nanoscale printing—working c̄ pharma companies for drug trials.
- Nano markets believe DPN can be a valuable tool for studying fundamental issues on colloid chemistry, surface science and nanobiotechnology.

Microarray technologies—are being developed for nanoscale bioanalysis—but due to limitations c̄ available microarray technologies—are not exploited for full range of drug delivery applications.

Nevertheless some limitations are being addressed at Nano level c̄ 'Nano arrays'—ultra miniaturised versions of the traditional microarray and can measure interactions between individual molecules c̄ resolutions as little as 1 nm.

↓

They are promised to revolutionise bioaffinity tests for proteins, nucleic acids and receptor—ligand pairs.

Quantum Dots (QDs)

- Is expected to take pharma industries by storm.
- QDs are semiconductor crystals that fluoresce when excited by a light source—emitting bright colours that can identify of track properties and processes in various biological applications.
- Have significant advantage over **traditional fluorophores** with reference to brightness, range of colours and their stability.
- QDs enable tagging variety of different biological components like proteins/different strands of DNA with specific colours.
- Thus QDs enable us to illuminate individual biological targets of choice *viz* genes, nucleic acids, proteins, small molecules, cancer cells or even tiny structures than these.
- 'Quantum Dot Corporation'—pioneering in use of QD with its extensive Q dot conjugate range.
- QD's use in Bioimaging is recognised by 'Science' Magazine as top 10 scientific break throughs of 2003.
- QDs can be applied to detect disease '*in vivo*' due to its specificity and high sensitivity.

  ↓

  Especially can be applied to detect cancer, where it is said it would take 4–5 years to detect initial cancerous cell growth, aiming to reduce this to 1 year c̄ help of QDs.
- QDs can also be employed to deliver drugs to specific sites, where ordinarily drug would not have access to deliver by itself.
- A company by name 'Advanced Magnetics' has a technology approved by FDA.

↓

This technology 'Combidex-Ferumoxtran-10' consists of iron oxide nanoparticles for use in MRI—to differentiate between cancerous and noncancerous lymph nodes—where combidex accumulates in noncancerous lymph node tissue—which enables doctors using MRI for different normal and diseased lymph nodes.

- Therefore drug industries are under tremendous pressure to develop new blockbuster drugs and nano markets believe $\bar{c}$ ↑ in understanding of nanotechnology and efficient application can lead to substantial new business revenues over next few years.

■ NOVEL DRUG DELIVERY SYSTEMS

- Efficacy of a drug is very much dependant on the method adapted for its delivery.
- Drug needs to be administered at optimum care because in case of **high doses** it may be **toxic** and in case of low dose **no therapeutic** effect at all.
- To minimize
  - (*a*) drug loss in form of degradation
  - (*b*) to prevent harmful side effects
  - (*c*) to ↑ drug bioavailability

Various drug delivery and drug targeting systems are currently under development.

Despite advances in modern medicine over last century—state of current medicine is limited and it is more of an art than science.

The current medicine approach fails to examine disease pathology at molecular level therefore still this so-called modern medicine is **Crude.**

- With nanomedicine—future/tomorrow's medicine/drugs the 'smart pharmaceuticals' shall have programable machines with a range of "sensory," "decision-making" and "effector" capabilities.

↓

They can check for overdosage before becoming active, thereby preventing accidental/intentional poisoning.

Nano based drugs can be totally "pure" and predictable in behaviour over their present day drugs.

Market for Advanced Drug Delivery

- Global market for advanced drug delivery € 37.9 billion–2000 and about € 75 billion in 2005.
- Controlled release € 19.8 B; Needless injection € 0.8 B.
- Injectable/implantable polymer systems € 5.4 B
  - —transdermal ∈ 9.6 B; € 12.0 B transnasal; € 17.0 B pulmonary
  - —€ 4.9 B transmucosal; € 0.9 B rectal; € 2.5 B liposomal
  - —€ 3.8 B genetherapy, Miscellaneous € 1.9 B.
- Developments occuring at rapid pace—as alternatives for injected macromolecules.
- Many novel drug delivery systems are being developed
  - — Soluble polymers
  - — Microparticles made of insoluble/biodegradable natural or synthetic polymers

— Microcapsules, cells, cell ghosts

— Lipoproteins, liposomes and micelles.

- Carriers can be made
  (*a*) Slowly degradable
  (*b*) Stimuli reactive *viz* pH or temperature sensitive
  (*c*) Targeted by binding them c̄ specific Abs against specific components of interest.
- Targeting by a drug molecule can be off **two** ways
  (*i*) **Passive—**Preferential accumulation of chemotherapeutic agents in solid tumors.
  (*ii*) **Active—**Surface functionalization of drug carriers c̄ ligands-selectively recognized by receptors on surface of the cells of interest—More precise targeting.

Nano Labelled Drug Delivery Systems

Injectable Delivery Systems

- Though injectable delivery systems have some discomfort, nano based injectable delivery systems provide improvements.

Example: Johnson and Johnson in 2005 announced 'Elan Nanocrystal Technology' in phase III clinical trial for injection of 'Paliperidone palmitate' an anti-schizophrenia drug—based on nanoformulation to older drug by overcoming original drugs insolubility by reducing to less than 200 nm of particle size.

↓

Results in extended product life cycle.

Implantable Delivery Systems

- Preferable to injectable drugs, as injectable drugs may cause toxicity if more than therapeutic range.
- Implantable DS gives scope for "time release" systems helping in minimizing peak plasma levels; reduce adverse reactions.
- Results in improving patient acceptance and compliance.
- They are expected to be used for delivering proteins, hormones, pain medications and other pharmaceutical components.
- p Sivida's has developed a **Biosilicon product**—it is a novel nanostructured material—stores active components in nanosized pockets—controllably release minute amounts of drug as **silicon** dissolves.

  ↓

  They are planning to develop for tissue engineering and ophthalmic delivery.
- 'Nano Shells'—promising approach for drug delivery in case of anti-cancer therapeutics.

↓

Have typically silicon lose, sealed in an outer metallic core.

By manipulating ratio of wall to core—they can be lined to scatter/absorb very specific wavelengths of light.

↓

Gold uncased nanoshells—used to convert light to heat—to destroy tumors by selectively binding to malignant cells.

■ ORAL DELIVERY SYSTEMS (ODS)

Many patients prefer ODS to implantables/injectables, therefore research to enhance traditional oral DS $\bar{c}$ help of nano-engineering improvements.

- Focus is to increase bioavailability.
- Treat previously untreatable conditions.
- Blood brain barrier (BBB) challenge—New chemical entities to treat brain disorders.
- 'Nano Pharm'—German Company—developed systems capable of reaching brain for anesthesia, cancer drugs and various other therapeutics.

Advantages

— No requirement to open BBB

— Potentially any drug can be used either hydrophillic or hydrophobic.

— Drug need not be modified.

- Researchers at University of Texas at Austin—Nanospheres for oral drug delivery. Nanospheres derived from hydrogels, highly stable organic compounds—swell when environment becomes acidic—successfully employed for formulating controlled release tablets and capsules—release drug only when hydrogel swells.

Topical Delivery Systems (TDS)

- Novavax—developed regulated topical systems—take advantage of properties of micellar nanoparticles.
- Used for hormone replacement therapy [Estrasorb and Androsorb].

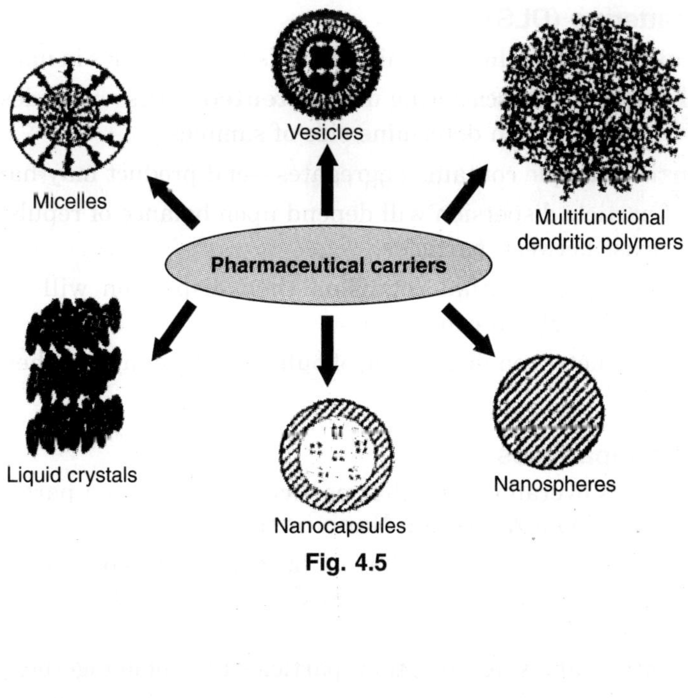

Fig. 4.5

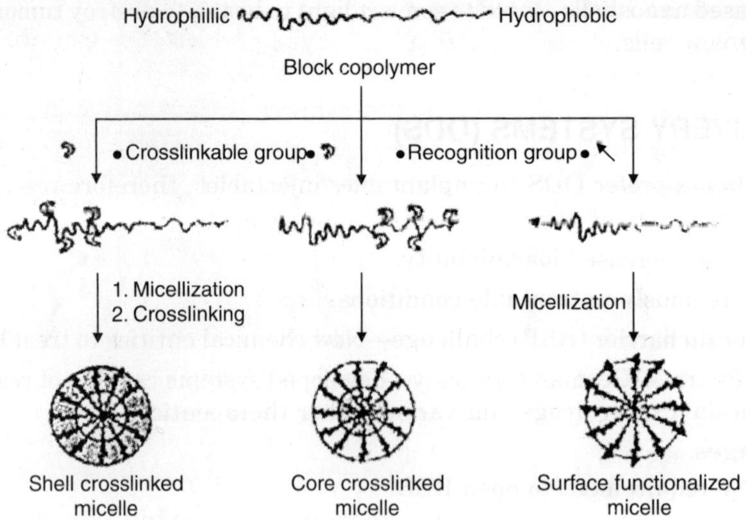

Fig. 4.6

RESEARCH INTO NANOPARTICLES

- Field of Nanoparticle covers wide range of interests *viz.*, chemistry, physics and materials science.
- Particle size, morphology and composition—manipulated to produce materials of different properties.
- Nanoparticles can increase the strength, hardness of metals and ceramics and protective coatings transparent.

Dynamic Light Scattering (DLS)

- DLS—ideally suited for the determination of size of particles in nanometer size range.
- Malvern Zetasizer Nanoscattering uses **patented** optics—provides exceptional level of sensitivity and one can determine size of samples.
- If nanoparticle sample contains aggregates—end product may have defects.

  Stability of particle dispersion will depend upon balance of repulsive and attractive forces that exist between particles.

  If all particles have mutual repulsion then dispersion will remain stable else Flocculation, Aggregation can set in.
- Malvern's Rheometers enable testing of bulk sample to say whether material is stable or not.

Zeta Potential of Nanoparticles

- Zeta potential is overall charge that particle accquires in a particular medium and can be measured on a **Zetasizer Nano** instrument.
- If all the particles in suspension have a large negative/positive zeta potential, then they will tend to repel each other and there is no tendency for particles to come together.
- If zeta potential values are low, then particles can come together and attract.

Why are Nanoparticles in the Headlines ?

- Companies are always exploring the ways by which they can use nanoparticles in their products.
- However applications have been limited because low availability of nanomaterials.

Reasons why companies are interested.

1. Competition
- Global market competition ↑, can products $\bar{c}$ nanoparticles give edge ?

2. Costs
- Can nano based products reduce costs ?

Nanoparticles can add value to product.

Example: Raw clay is $ 20 per tonne while nano particulate clay $ 2,000 per tonne.

By 2013–5,00,000 tonnes demand for nano particulate tonne.

3. Technologies
- Low cost digital cameras

4. R & D
- Expensive process: Governments of all industrial countries are sponsoring schemes to address this issue by facilitating projects between nanoparticles, manufacturers, industry and academia.

5. Health and Safety
- Issue of nanoparticles escaping into the environment.
- Asbestos scares—Blue asbestos only other in domestics not of great concern.
- Full understanding is required and regulations for handling them to be drafted carefully.

Physics of Nanoshells

- They are new type of optically tunable nanoparticle composed of dielectric 'Silica' Core coated with an ultra-thin metal like 'Gold'.
- Gold nanoshells possess physical properties similar to gold colloid, show strong absorption due to collective electronic responses of metal to light.

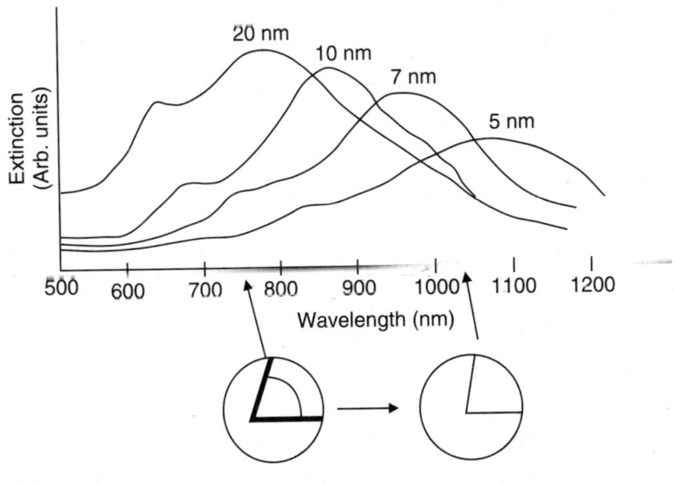

Mic scattering depicts nanoshell plasmon resonance wavelength shift as a function of nanoshell composition $\bar{c}$ 100 nm core.

Fig. 4.7

- This optical absorption of nanoshells—yield a brilliant red colour, has high potentiality in medical products like pregnancy testing kits.
- Optical response of gold nanoshells depends dramatically on the relative sizes of the nanoparticle core and thickness of gold shell.

↓

- By varying relative core and shell thickness—colour of gold nanoshells can yield broad range of optical spectrum—spanning visible to near IR spectral region.
- Ability to "tune" nanoshells to a desired wavelength is critical to '*in vivo*' therapeutic applications.
- Human blood and tissue minimally absorb certain near IR wavelengths of light, enabling us to use an external laser to deliver light to nanoshells either in a tumor (for thermal destruction/imaging), a wound (for wound closure or tissue repair) or whole blood—to diagnose disease.

BUCKYBALLS AND BUCKYTUBES

Our understanding of C60 wouldn't be, if it weren't for astronomy. Harold W. Kroto was analyzing radio astronomy patterns, when he discovered a naturally occurring unknown carbon in interstellar dust. Now, they are believed to be throughout the galaxy, and possibly surrounding (in) "Red Giants" and "Black Holes". Black holes, black body radiation, cosmic strings, fullerenes, structure, properties, and applications seem to be linked. I believe they are all intricately connected and have touched on them in my report.

Stephan Hawking's theories state that there is a form of blackbody radiation which black holes produce. This happens when subatomic particles from 'sets' of two, next to the balck hole. One is sucked in, and one radiates off. Stephan Hawking says that if enough move off, the black hole disappears. The energy for the particle to radiate off, is taken from the black hole, and therefore diminishes its mass. Since a black hole's size has a relationship to "how many molecules are in it", every time it loses mass, it shrinks. It appears that this "black body radiation" characteristic of black holes, is also a characteristic of buckyballs. Apparently fullerenes also can absorb and produce black body radiation. "Cosmic strings" attached to the black hole, are another "energy exit." With this being said may be black hole aren't the big vacuum in space they were thought to be.

The Origins of Life

Scientists have found gas filled buckyballs embedded in 4.6-billion-year-old meterorites. These rare space gases are similar to gases, which were present when the solar system was first forming. If buckyballs have truly come from another planet, then it could mean they have the durability to survive the trip through the universe. This finding supports the theory that atmospheric gases and organic compounds arrived on earth during asteroid and comet strikes, early in the planet's history when impacts were numerous. Scientists have found buckyballs in large amounts embedded in several meteorites. They hypothesize that the key ingredients for life on earth were made from these gases. If this is true it can lead to further speculations that we are not alone in the universe.

Buckyballs

First noted in the stars and later identified in common 'soot', fullerenes are now speculated to be part of the cosmic molecular streamers stretching to the center of the galaxy.

C60 was only produced mechanically in 1990. This was after an experiment done by a scientists trying to recreate red giant star conditions. They realized that they had found a new type of carbon, which we know as C60 (other common terms for this are fullerenes and buckyballs).

Natural carbon can exist in numerous forms. Most people are familiar with graphite and diamond, but there is another form—fullerenes. Often they are mistakenly called new, but fullerenes have been discovered to exist in interstellar dust as well as in geological formations on Earth. Fullerenes are big carbon-cage molecules. The most common one is C60—also known as a "buckyball"—but some other relatively common ones are C70, C76, and C84. Fullerene cages are about a billionth of a meter. In atomic terms, they are enormous. But fullerenes are still small compared to many organic molecules. C60 is made of 60 carbon atoms. All of which are linked together in a pentagon and hexagonal pattern. In fact, it has the identical shape of a soccer ball. With this being said may be one of the most monumental breakthroughs have been right under our noses, or right under our feet, the whole time.

Buckytubes

In 1991, Sumio Iijima discovered buckytubes. He discovered them while examining buckyballs. He noticed the walls of the tube consist of carbon atoms arranged in hexagonal patterns with one pentagon to cork each end. Capping of the tube requires that the walls bend in a 3-dimensional manner and this is achieved by the incorporation of 5-membered rings in the carbon framework. Buckytubes are super fine filaments made of hexagonal "chicken wire" carbon mesh. Both buckytubes and buckyballs are from carbon, however, as the "balls" form, they incorporate the six and five sided carbons throughout the "ball". As the "tube" rolls up, it usually only incorporates the five-sided carbon at the ends of the tube. However, having said that, the tube can be "encouraged" to incorporate the five-sided carbon elsewhere in the tube as well, and this considerably changes the tube's abilities and properties.

Buckytubes are extremely strong and have high tensile strength. Their strength is equal to a diamond, the hardest substance known. They are unique not only in the way that they can conduct electricity, but also react to it. When a carbon nano tube is charged with electricity, it causes it to move up and down in a wave-like pattern. These facts make it perfect for a number of uses. One of uses being considered, is tethering satellites to the earth with a 22,000-mile buckytube rope. This would make it easy to take down and put up satellites in space. Buckyballs and buckytubes are having new and amazing properties discovered everyday. How much they capable of performing and transforming in the pressure and temperature variances of outer space, in proximity to Red Giants, space anomalies, and black holes is yet to be discovered.

These tubes have many properties which will be useful in their development. (*a*) Electrical conductivity: so far they have been found to be the best conductor of electricity on a nano scale level. (*b*) Thermal conductivity: they have been shown to be as good as diamonds. (*c*) Mechanical: they are the strongest and toughest fiber that has been found yet. With increased research and methods, we really might be able to tether satellites to the earth, like helium balloons. (*d*) Carbon based: they are carbon based, so it is something many are already familiar with. Carbon is the basis of most materials we use everyday. (*e*) Molecular perfection: they have a high degree of uniformity and freedom from defects. When there is a "defect" it always seems to end up being another amazing property, instead of a problem. (*f*) Self-assembly: because of the strong forces between them, they can "automatically" group together, in some situations.

Last year only 3 kilograms of nano tubes were produced, selling for about $ 300 per gram. This is about 30 times the cost of gold. Analysts estimate that costs will soon be down to 50 or 60 dollars per gram. Companies such as IBM, Xerox, Mitsbushi and DuPont have already registered patents on buckyballs, and are currently experimenting with new concepts.

These tiny paradoxes are small enough to migrate through cell walls, conduct electricity as well as copper, conduct heat as well as diamond, and are 100 times stronger than steel at one-sixth the weight. As well, they have fluorescence. That is, they can absorb one wavelength of light, and emit another wavelength. One way this could be useful is in the fiber optics industry, medical industry and oilfield. As well, different diameters of the buckytubes, absorb and give off different colours of light. A grouping of different sized diameters could be used to diagnose more than one medical condition at a time, or, a grouping of different sized diameters could be injected into an oil well. When they were "circulated" back out, the colour changes could give engineers critical information about the conditions downhole, as well as what is there. These and other applications have been expanded on ahead.

Applications

Buckyballs and buckytubes appear to be the Industrial Revolution of this century. They entice big businesses, defense/offense, humanitarians, individuals and the masses with their possibilities and promise. Having "come from the stars" (and possibly black holes/and/or a parallel universe) matters not at all to us on earth. We seek to "make them our own."

Although the following examples are separated into a few categories, almost any of the applications could be developed to apply to any circumstances conceivable.

Space Applications

Space exploration has always been difficult due to the elements and other harsh factors. Buckytubes and buckyballs work well under ANY temperature variations making them perfect for space.

1. NASA Ames Research Center has the following amongst it interests: Chemical storage of data; Micro Electrical Mechanical Systems (MEMS) gears and pumps; Nanotube heterojunctions; Nanotube as probes for metrology and measurement; Nanotube amplifiers; Nanotube robotic system manipulators; Molecular bearings; Nano capillaries; Nanotube ferromagnets; Coaxial nano cables; C60 coatings for magnetic disks.
2. Electrochemical, strength, communications, shielding, and optics are a few of the applications being pursued by NASA.
3. Even improved lighting in the spacecraft might benefit.
4. Biosensors for monitoring space crew, and/or space environments.
5. Protective coatings on any surface, which might be exposed during space flight.

Applications in the Oil Field

One major concern with oil companies is the wear and tear of their equipment in harsh conditions like in Antarctica, where the temperature is below freezing. As many people know machines do not operate well under cold conditions. The gears start to freeze up, and before you know it the whole machine cannot move. Studies have shown when engines are coated with buckyballs it decreases the engines' wear and tear tremendously. In fact some studies have shown it works so well that no visible wear appears. The added benefit is they work in any temperature. Which would be perfect for machines in Antarctica. If oil companies were to coat their machines in buckyball mixtures, they would have no problem with the gears freezing and wouldn't have to fix the equipment as often. This technology could also be applied to drill bits increasing their strength and performance.

1. Safety is critical in oilfield work, as elsewhere. Accidents cost misery, time off work, and of course money. Some situations have hazardous gases which have killed many oilfield workers. Now, with Buckytubes/Buckyballs, there is an opportunity for a solution. A silicon wafer, with a single embedded buckytube, can turn that wafer into a gas sensor.
2. Because buckytubes fluoresce, and different sized buckytubes absorb and give off different light ranges, they could have applications in the oilfield. A grouping of different sized diameters could be injected into an oil well. When they were "circulated" back out, the color changes could give geologists and engineers critical information about the conditions downhole, as well as what is actually there.
3. As well, this property of fluorescence could be used extensively in oil explorations. Mixtures of buckytubes could be injected in all sorts of locations (many which people would want to go to, or be easily able to go to), and the changes in light emitted could be read by satellites.
4. Buckytube transistors, as well, could make the equipment needed to explore for oil, and manage oilfields, much smaller. With enough commercialization they will become cheap, as well as small. Much the way the price of calculators has dropped over the last thirty years. At least some of the awkward heavy equipment could be done away with. IBM, AT and T, Mitsubishi have both been leaders in investigating this. The advantage will be circuits that are smaller, cheaper and faster.
5. A significant expense, both for replacing parts and loss of work time, in the oilfield is wear and tear on the actual parts of the drills, pump jacks, gears, and various other moving parts involved in oil exploration, drilling and production. In very cold regions (Artic), even getting to the sites can be difficult because the gears don't move well against each other in cold. There are two ways buckyballs could solve these problems. (*a*) Moving parts can be made with coatings of the buckyballs embedded in them. The tests have shown virtually no wear in ANY temperature. (*b*) As well, another company has developed an oil with a buckyball component. They have run the car engines with this oil in, to coat the engine parts. Then they drained the oil. Again, tests found little or no wear on moving parts after running them great distances. It seems that the protective component of buckyballs adheres well to metal either way.

Medical Applications

Buckyballs are making quite a stir in the medical world too. Their ability to trap gasses such as helium and store them, makes them suitable candidates for chemical and biological sponges. For example, when the brain is injured it releases deadly nerve toxins, which can have devastating effects on the brain. Doctors could use them as sponges to soak up all the harmful chemicals surrounding the brain. Then the chemicals would harmlessly pass through the body, without damaging any organs.

1. Researchers have found that buckyballs are a close fit on the site to block the HIV molecules. To put it in a more colourful way, they have found they can stop HIV activity by "plugging in" a buckyball. If this is successful, the possibilities are limitless for other similar applications. Further, since drugs or treatments (*i.e.*, radiation) can actually be put inside a buckyball, it could be used in this way for cancer treatments.

2. Other researchers, as well, have found ways to alter buckyballs to fight diseases. They believe they can tailor make them to fight bacteria and fungus.
3. Researchers have developed buckytube artificial muscles which appear to be stronger and more durable than actual muscle. Possibilities for using this to repair human problems are exciting (**example:** heart repair, other muscle repair in the body, etc.).
4. Columbia University researchers are working on ways to coat buckytubes with protein, and treat in such a way that the coated buckytube fluoresces in the presence of certain other chemicals or "triggers". The person could then be "scanned" and determinations could be made about levels of chemicals in the body (calcium, lead, etc.) as well for diagnosing diseases (cancer, make it specific to different viruses, etc.).

   http://www.nature.com/nsu/011004/011004-2.html
5. Medical sponges is yet another use. Buckyballs could soak up free radicals in the brain, which are harmful chemicals the body produces after an injury.
6. Individual buckytubes have been used as tips on scanning-force microscopes. They have been found to be more durable than other tips previously used.
7. Individual buckyballs have successfully been used as part of VERY miniature circuits. This may lead to "bucky circuit inserts" which could be used to help "repair" damaged circuits in the brain or spinal cord.
8. Their excellent properties as lubricants, could also make them useful to insert into damaged joints where natural lubrication has eroded.
9. They are in use as a coating for a special type of lung X-ray.
10. Methods are being developed to utilize them as filters in areas such as organ transplants.
11. C60 is being developed by many different programs to treat a wide variety of medical concerns. One company, alone, currently lists seven significant medical uses. (*a*) C60 photo therapy for cancer (*b*) HIV treatments (*c*) use as antibodies (*d*) use in treatment of diseases such as Alzheimer's, Parkinson's or Lou Gehrig's Disease (*e*) designing C60 to reach specific tissues, for example to "highlight" certain tissues or organs during X-rays (bones, liver etc.), or to deliver drugs or treatments to specific areas. (*f*) improved materials for X-rays (*g*) drugs specific for treatment of osteoporosis (*h*) coronary applications such as stents or replacement of damaged valves.
12. Buckytubes have been found to be a possible replacement for any narrow "tube-like" part of the body *i.e.*, veins, neurons, muscle.
13. Weighing viruses, possibly to help identify their properties, is now also possible.

Military Applications

Buckytubes also have a wide variety of military applications, such as subs. If the hull of the sub was made of nano tubes it would be virtually indestructible. Plus instead of using propellers to push the sub forward, you could charge the hull of the sub with quick short bursts of electricity. This would cause the hull to move back and forth like a shark propelling it forward. Making virtually no noise in the water, as opposed to the conventional sub, which uses propellers and can be heard would make it undetectable. Oil transports would be virtually indestructible making it impossible for them to spill or leak harmful pollutants into the water.

Civilization anticipates endless possibilities and applications for buckytubes. Perhaps the most obvious and most practical use would be putting its toughness and durability to use.

These qualities could be utilized in almost any area where strength is desired. For instance, tanks and other machines that demand superior armor would be a perfect candidate. Imagine a light weight tank that weighed only two cars (and was just as fast as one) plus 100 times the strength that steel offers. This would mean it could withstand a mine or missile attack without being scratched.

Another important development is manufacturing circuits on the molecular level. This could mean supercomputers in the palm of your hand, an entire CIA department could fit into your backpack. A spy could have everything and anything right there. Conventional gadgets like tape recorders will be reduced to the size of a pencil lead, and able to store thousands of hours of audio. Eventually even the spies will be replaced with tiny robots, the size of a bread crumb, but with intelligence, and entire array of tools at its disposal.

1. Buckminster fullerene (C60) stops light—the brighter the light the more effectively it stops it. Nano and pico-second laser pulses are effectively and instantly opaque by small quantities of C60. A helmet visor treated with fullerene will instantly block an incoming laser beam—the stronger the ray, the faster the glass turns dark. This feature could be developed to "shield" or "cloak" military planes and vehicles.
2. Also, since many bombs and missiles are laser (light) guided, planes, vehicles, buildings, etc., could be coated in this (it is becoming cheap enough this is actually possible), and it would block the "lock on light".
3. In many military operations, planes, helicopters, and other machines are needed in places where it is not possible or easy to do repairs. As well, some locations have considerable problems contributing to the wear and tear on moving parts. There are two ways buckyballs could solve these problems. (*a*) Moving parts can be made with coatings of the buckyballs embedded in them. The tests have shown virtually no wear in ANY temperature. (*b*) As well, another company has developed an oil with a buckyball component. They have run the car engines with this oil in, to coat the engine parts. Then they drained the oil. Again, tests found little or no wear on moving parts after running them great distances. It seems that the protective component of buckyballs adheres well to metal either way.
4. Buckyballs have "black body radiation" properties: a black body absorbs and re-emits all radiation landing on it. This aspect, as well, could be exceptionally useful in "protective" or "stealth" military aspects.

Wrap-up

Buckyballs and buckytubes seem to be a case then, of if you can visualize it, someone can develop it. They do seem to have been created near (may be in?) black holes. Perhaps in our thirst and hunger for new benefits, we'll be able to use them to find an answer to where life began; are there other universes out there; is there a way through a black hole; or if the "seed" for human life came from space (possibly in buckyballs) how did it get to "space" ? Enhanced health, technology, leisure, and defense opportunities are worth exploring and worth enjoying. However, with any luck we won't exploit them largely to "build better bombs", or to "build a better mouse trap".

5

Oxide Nanoprecursors: A Technological Perspective

Way back in 1959, famous scientist Richard Feynman was able to visualize the properties of nanoparticles. He had predicted that the reduction in size of a substance to atomic scale would lead to materials with extremely new properties. Scientists experienced the truth in his prediction when they could actually bring the size of the material down to nanoscale and develop different techniques for characterization. Research activities in nanomaterials are two fold. Scientists have studied technological advantages of using precursors in finite size range. At the same time rich and fascinating physics of fine particles has attracted the research workers.

After studying the physics of different oxide fine particles, we have established a general rule. According to our observation, with the reduction in size to nanometer scale, the lattice of oxide material tends to acquire higher structural symmetry. As a result, the substance exhibits altogether new properties since there is always a structure-property co-relationship. However, this paper deals with the technological advantages of nanoparticles, which we have come across during our study in past few years. With the reduction in particle size obviously there is increase in specific surface area of the sample. Reduction in size also enhances the ability of particles to cover large surface more uniformly. Fine particles therefore find applications in catalysis, paints and pigments, magnetic recording etc. Rate of sintering is inversely proportional to initial particle size. As a result, fine particles are more effective when used as precursors to obtain highly dense products, which can even replace single crystal.

In this paper, the aim is to compile the results obtained by us while studying a variety of oxide systems. It would certainly help to build an overall picture about the technological potential of nanoparticles. The paper is divided into 4 sections. Section 1 consists of different routes developed and used for synthesis of oxide fine particles. In Section 2, techniques used for size determination are mentioned. Compiled results obtained on different oxide systems are included in Section 3 while Section 4 gives conclusions.

■ SYNTHESIS OF OXIDE NANOPARTICLES

Methods used for the synthesis can be categorized into three types:

(*i*) Mechanical: Ball milling

(*ii*) Physical: Solid-solid phase transition, Spray drying, Liquid drying

(*iii*) Chemical: Co-precipitation, Sol-gel, and Micro-emulsion

Each technique has its own figure of merit and the choice of the technique depends upon the material of interest. However, it has been observed that the particles derived through chemical route have controlled chemical homogeneity, stoichiometry and size distribution. In case of multi-cation systems, it is possible to avoid solid state diffusion step since the complex formed in wet chemical route directly leads to reacted phase on calcination. It helps to reduce the calcination temperature and hence to control the particle size.

Mechanical Method

Ball Milling

It is the simplest and easiest method to achieve particles at sub micron level. High purity oxide materials, in the proportion required in the final product, are mixed together. Mixing is often performed by wet milling in a rubber-lined pot using stainless steel or alumina balls. The mixture is then dried and calcined. The hard cake obtained during calcination process is ball milled once again for a long time to get fine particles. Though process is simple it has some serious limitations such as, lack of control on chemical purity, homogeneity and particle size distribution. The process is lengthy and takes about a month to bring the particle size to sub micron level.

Physical Methods

Solid-solid Phase Transition

Palkar *et al.* were the first to show that the process of phase transformation could be effectively used to synthesize nanoparticles. It was observed, that, irreversible phase transformation is accompanied by marked reduction in particle size. The reduction in particle size is proportional to change in unit cell volume at transition. In order to control the particle size, the calcination temperature needs to be restricted just above the phase transformation. It helps to avoid the growth of the freshly nucleated phase. The method is applicable if suitable metastable polymorph is available. In certain cases it could be an alternative to wet chemical route.

Spray Drying

Fig. 5.1 shows the set up used for spray drying. In this method, the solution to be dried is fed to the eye of the centrifugal atomizer disc. Due to rotation of the disc at high speed the solution is thrown out in the form of a fine mist. Hot dried air is simultaneously introduced through an annular opening into the drying chamber. This hot air mixes continuously with the mist of atomized liquid and an instantaneous evaporation of the liquid takes place. The non-volatile part is left in the form of dry powder. The powder falls down towards the outlet aperture in the conical bottom of the chamber and from there it is carried into the dynamic cyclone separator by the air current. The powder falls down into a container and the drying air is discharged from the top of the cyclone. The dry powder is then calcined to get the required phase. In this technique, particle size is controlled by speed of the disc, solution concentration and calcination temperature. Special care is required to preserve the particle size if the material is hygroscopic. In case of multi-cation systems, to obtain reacted phase, the reactants have to undergo solid-state diffusion since there is no complex formation.

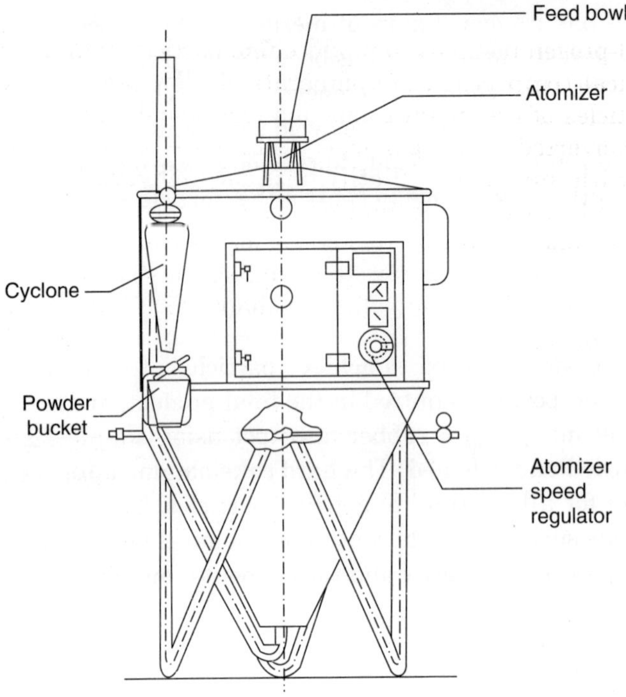

Fig. 5.1 Spray dryer

Liquid Drying

Liquid drying of an aqueous solution of metal salt requires in general, a drying liquid having a high degree of solubility for water but not for metal ions. Since the interest is in getting oxide phase at final stage, the soluble salts like acetate, citrate etc. are commonly used. In this method, the stoichiometric aqueous solution is sprayed as a broken stream of droplets into the vortex of swirling bath of hygroscopic liquid. Acetone is used as a hygroscopic liquid in most of the experiments. The ratio of the total volume of acetone to the salt concentration is ~ 10 : 1. Under these conditions nucleation of particles is essentially instantaneous. Acetone-water mixture is removed from the dry powder by vacuum filtration. Since no formation of complex occurs, in multi-cation systems solid-state diffusion process is unavailable to achieve the reacted phase. Particle size is decided by the solution concentration, and calcination temperature.

Chemical Methods

Co-precipitation

The lengthy ball-milling process involved in mixing the constituents of multi-cation system could be avoided in co-precipitation route. The constituents are weighed in stoichiometric proportion and dissolved to get the stock solution. All the cations are then co-precipitated to achieve the complex formation by selecting suitable precipitating agent and pH. In case of oxides the precipitate could be in the chemical form of hydroxide, carbonate, oxalate etc., so that after decomposition it converts to oxide. In this process the particle size is determined by the solution concentration, rate of precipitation and calcination temperature. However, during the drying of the precipitate, there is a possibility of agglomeration of the particles and hard cake obtained in this step needs ball milling.

Sol-gel

This is one of the well-proven methods to produce fine particles with narrow size distribution and controlled chemical composition at comparatively low temperatures. In this process, dispersion of the particles of the metal compound (sol) usually in an aqueous phase is first prepared and then converted to gel particles. Gelation is achieved by means of chemical dehydrating agent with surfactant. 2-ethylhexanol (dehydrating agent) and SPAN-80 (surfactant) is commonly used in the process. In principle, particle size achieved at the precipitation state is maintained during sol and gel formation. Since agglomeration is avoided the process gives better control on particle size and size distribution. Column technology is used, if geleted material is required in spherical shape (Fig. 5.2). The gel is then dried and calcined to get reacted product.

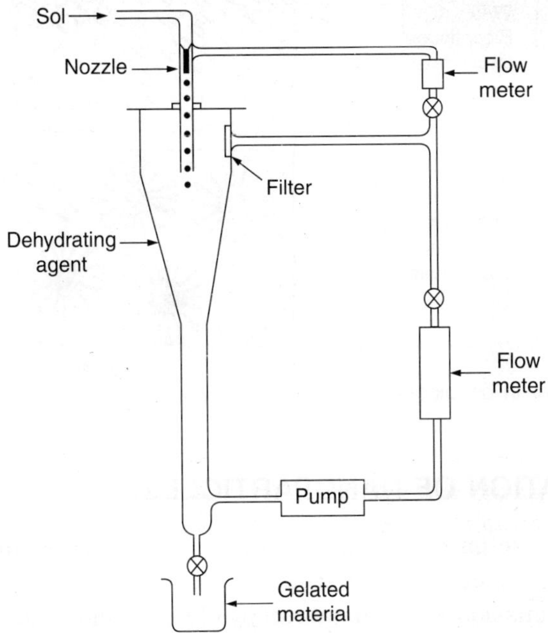

Fig. 5.2 Glass column used in sol-gel process to produce geleted material in spherical shape

Micro-emulsion

When two immiscible liquids are mixed together, emulsion is formed. However, to get micro-emulsion (stable and transparent system) the globule size of the dispersed phase needs to be less than 1000 Å. Micro-emulsion is basically a three phase system *i.e.* Oil (organic phase), Water (inorganic phase) and surfactant (cationic, anionic or non-ionic). Surfactant remains at the interface and helps to prevent coagulation of dispersed globules. Different types of micro-emulsions (Water dispersed in oil (W/O $<<$ 1), oil dispersed in water (W/O $>>$ 1) or bicontinuous structure. (W/O ~ 1)) are shown in Fig. 5.3. The method mainly consists of the following steps,

(*i*) formation of micro-emulsion with proper choice of three phases

(*ii*) precipitation of inorganic phase

(*iii*) separation of precipitate and

(*iv*) drying and calcination.

The particle size is mainly controlled by the solution concentration and calcination temperature. Palker *et al*. were the first to show that micro-emulsion could be used to synthesize oxide fine particles.

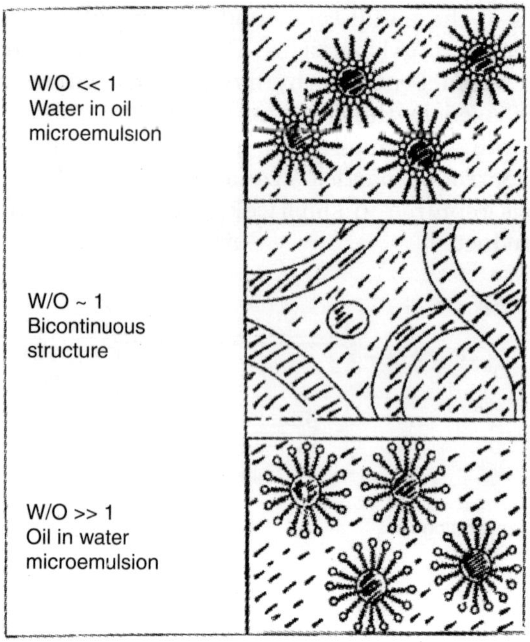

Fig. 5.3 Water in oil, bicontinuous structures and oil in water microemulsions

SIZE DETERMINATION OF NANOPARTICLES

After the synthesis of phase pure sample, it is important to determine particle size and size distribution before carrying out further studies. X-ray line broadening, scanning electron microscopy (SEM), transmission electron microcopy (TEM) and surface area analysis are some of the techniques that are commonly used for the purpose. Though there are various techniques available for size determination, no technique is fool proof. Size calculated from X-ray line broadening using Scherrer's formula, is actually the size of the coherently diffracting domains. Unless the particles are single domain, the size determined by X-ray is always smaller than that obtained by other techniques. It is not possible to find out the size distribution using X-ray line broadening. SEM and TEM could be used to determine size as well as size distribution. However, during sample preparation bigger particles tend to settle at the bottom of the container. Hence, the sample picked up for the measurement, does not always give true representation of the particle size. This can lead to error in measurements. Size calculated from surface area, actually gives equivalent spherical diameter. It is assumed, that, all particles are spherical and there is only monolayer adsorption of gas during surface area measurements. Under the circumstances, there are chances of error in size determination. Moreover, the technique cannot be used to estimate the size distribution.

RESULTS

In order to reveal the technological ability of nanoparticles, results obtained on different oxide (structural ceramic, high T_c oxide superconductor, electronic ceramics) samples are presented in this section.

Efficiency of sol-gel derived nanostructured γ-alumina porous spheres as an adsorbent in liquid chromatography

Alumina (Al_2O_3) exists in several structural phases. γ-Al_2O_3 is the low temperature phase with crystalline lattice closely related to the spinel structure. The unit cell of γ-Al_2O_3 is formed by cubic closed packing. The octahedral aluminium sites are fully occupied while vacant sites are randomly distributed over the tetrahedral interstices. These vacant sites in the structure are mainly responsible for the adsorbent nature of γ-Al_2O_3. The adsorption capacity of γ-Al_2O_3 is mainly used in liquid and gas chromatography to separate chemical compounds with different adsorption coefficients.

Efficient performance of the adsorbent, in terms of clean separation, could be obtained by increasing the specific surface area (SSA). In order to achieve enhancement in SSA particle size of the adsorbent needs to be reduced. It is but natural, that, the chromatographic column prepared by making use of fine particles gets densely packed. It creates obstruction in the flow of the eluting agent used in liquid chromatography. It eventually results into an increase in time required for the separation process. For on line separations in industry, one needs to minimize the required time without compromising with the purity of the separated product. This problem could be solved by using highly porous fine particle.

The results obtained by using γ-Al_2O_3 obtained through various sources for purifying tetraphylporphyrin (TPP) and its *n*-methylated derivatives are shown in Table 5.1 and Fig. 5.4. The experimental details of the novel sol-gel process developed, to obtain ~0.5 mm diameter porous γ-Al_2O_3 spheres composed of 4-5 nm size particles with their characterization, are reported elsewhere.

Table 5.1 reveals that sol-gel derived alumina has very high specific surface area and hence a very high adsorption capacity among all tested samples. High surface area coupled with high porosity helps to achieve improved performance and desired economy in processing time when used as an adsorbent in liquid chromatography (Fig. 5.4). This superior behaviour could be highly useful during on line separation in chemical industry.

Table 5.1. Surface area, pore volume, adsorption activity and time required for chromatographic separation for γ-Al_2O_3 obtained from different sources

Sample	*Surface area* (m^2/g)	*Pore volume* (cc/g)	*Adsorption activity* (moles/g)	*Time required* (h)
Sol-gel	300	1.28	2.9	9
Fluka	135.8	0.386	0.9	22
BDH	80	0.28	0.1	27

Efficiency of Pd Impregnated sol-gel derived γ-alumina porous spheres as catalyst

The study of catalysts is essentially a study of the surface chemistry. The efficiency of the catalyst is in its interaction with the environment. In surface assembled metal catalyst, metal is distributed in the form of fine particles on support material. The classification of support material is done in terms of porosity and surface area. An enormous increase in activity could be obtained if an appropriate support material could be synthesized. Generally, the metals like ruthenium, rhodium, palladium, silver, platinum and gold are used as catalysts while alumina, silica, magnetia etc. are used as supports. The supported catalysts have several advantages over simple catalyst like, large area of metal is exposed, sintering of metal is reduced and the surface accessibility of the catalyst to reactant is improved. The heat produced in an exethermic reaction is dissipated. The approach allows the controlled use of expensive materials, since they are more or less confined to the surface of the support.

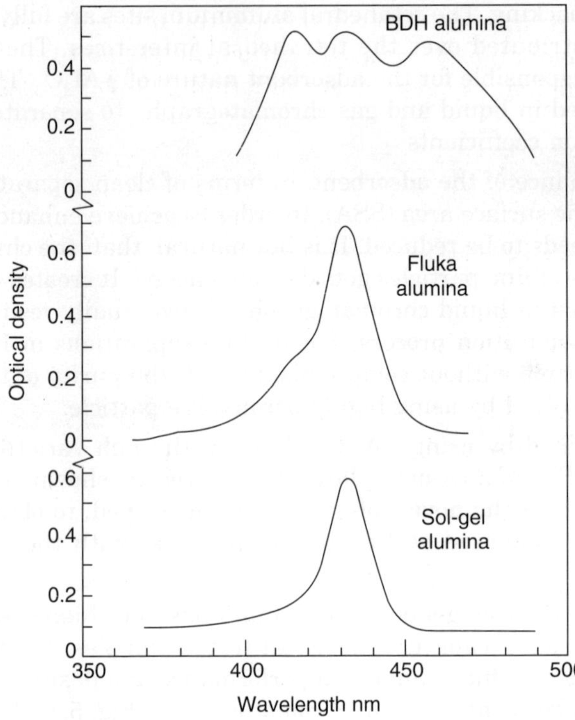

Fig. 5.4 Absorption spectra of TPP and its *n*-methylated derivatives

The activity of a supported catalyst for a given reaction should be referred to as the unit area of metal surface exposed to reactants. The activity per unit area of accessible surface is called the surface activity. The rapid ability of the catalyst to interact with a reacting system to give a certain chemical composition and structure is desirable. Porous γ-alumina spheres synthesized through novel sol-gel process, were tested as catalyst support after impregnating with palladium. The results were compared with the commercially available support material but impregnated with Pd in our laboratory, under identical conditions. Hydrogen-oxygen combination reaction was used to determine the catalytic behaviour. Catalytic efficiency was calculated from the percentage of the unreacted gases. Table 5.2 clearly indicates, that, though both alumina samples are soaked under identical conditions, the amount of Pd impregnated on sol-gel alumina is more. This could be attributed to higher adsorption activity of the material. The specific surface area of the sol gel alumina is almost 4 times to that of commercially obtained alumina. Hence, Pd on sol-gel alumina is distributed over larger surface. With the result, that, larger surface of Pd is available for the catalytic reaction. γ-alumina is

Table 5.2. Catalytic efficiency of Pd impregnated alumina

Catalyst support	***Surface area of support***	***Pd/100g of support***	***Catalytic efficiency***
Sol-gel alumina	300 m^2g^{-1}	0.698	100%
Commercial alumina	80 m^2g^{-1}	0.50	93%

a known adsorbent material. Hence, it is more likely that during soaking Pd remains adsorbed on the surface rather than getting absorbed. SEM pictures also indicate uniform coating of Pd on the surface (Fig. 5.5). Equivalent spherical diameter as calculated from specific surface area indicated that Pd, particle size could be of the order of 5 nm. Enhanced surface of nanostructured Pd, available for catalytic action, helps improve the catalytic behaviour. Spherical shape of alumina is advantageous in forming uniform packing of the catalyst bed.

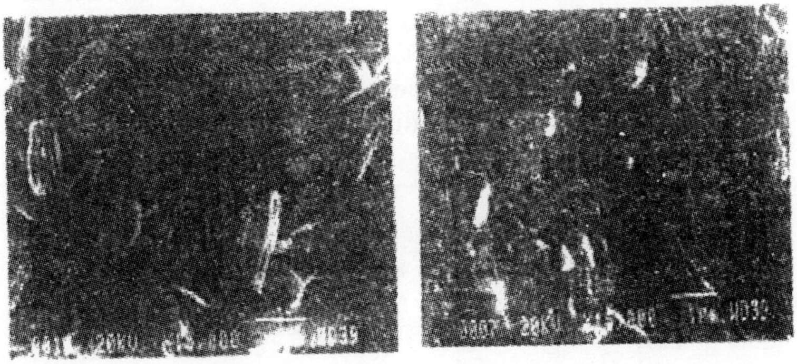

Fig. 5.5 SEM picture of γ-alumina and Pd coated γ-alumina porous spheres derived through sol-gel process

Synthesis of high T_c superconductor $YBa_2Cu_4O_8$ (1-2-4) under the condition of normal oxygen pressure

The superconductor (1-2-4) was originally observed as a lattice defect structure and then as an ordered defect structure in microstructural studies of $YBa_2Cu_3O_7$. Synthesis of (1-2-4) in bulk form was first reported by Karpinski *et al*. using high oxygen pressure (400 bar) and high temperature (1040 °C). Cava *et al*. reported, a two-step method, using alkali carbonate catalyst. However, no technique could produce phase pure (1-2-4) sample. Studies on (1-2-4) system had indicated, that, it might turn out to have several advantages over other ceramic superconductors discovered till then. It was therefore interesting to find out a simplified method that could produce single phase (1-2-4) material.

It was seen that the sol-gel route could be used to produce single-phase (1-2-4) sample without any requirement of high oxygen pressure. The hydroxy carbonate complex was obtained from stoichiometric solution of the constituents by using sodium carbonate as the precipitating agent. The particle size of the order of 4-5 nm obtained in precipitation step was maintained during the sol and gel formation. The geletion was carried out using chemical dehydrating agent (2-ethyl hexanol) along with surfactant (SPAN – 80). Single step decomposition of the complex (900 °C/30 min) coupled with an enhanced oxygen uptake due to increased specific surface area, facilitated the formation of the (1-2-4) phase during calcination from the oxygen rich atmosphere. For the first time synthesis of phase pure (1-2-4) has been reported.

Densification of Lead Zirconium Titanate (PZT)

PZT ceramic is well known for its best piezoelectric properties. It is, therefore, used in a variety of sensors and actuators, MEMs etc. The piezoelectric properties of PZT are very sensitive to composition, grain size and density of the sample. If the grains are too small the ferroelectric polarization is 'locked-in' and is difficult to realign by an electric field. To get optimum properties the ceramic should be highly dense (+95% of theoretical value) with grain size of the order of 1 micron. The difficulty in sintering of PZT with desired density and grain size arises due to volatile constituents like PbO. If the temperature is raised during sintering, there are PbO losses. PbO losses affect the stoichiometry of PZT, which ultimately impairs piezoelectric properties. Highly sintered and stoichiometric PZT in pellet form is also required as a target for pulsed laser deposition to get good quality PZT thin films. PZT thin films have variety of important applications including non-volatile memories. It was observed, that, by using sol-gel derived PZT precursor (~16 nm in size), it was possible to bring down the sintering temperature to 1200 °C from 1400 °C (required for PZT obtained by conventional route). The reduction in sintering temperature helped to minimize PbO losses and maintain the stoichiometry. The highly dense product (+99%) could withstand poling voltage of the order of 70 kV/cm. Ultimately it could give d_{33} coefficient (piezoelectric coupling factor) of the order of 293×10^{-12} Coul. Newton^{-1} (highest reported value) as seen in Fig. 5.6.

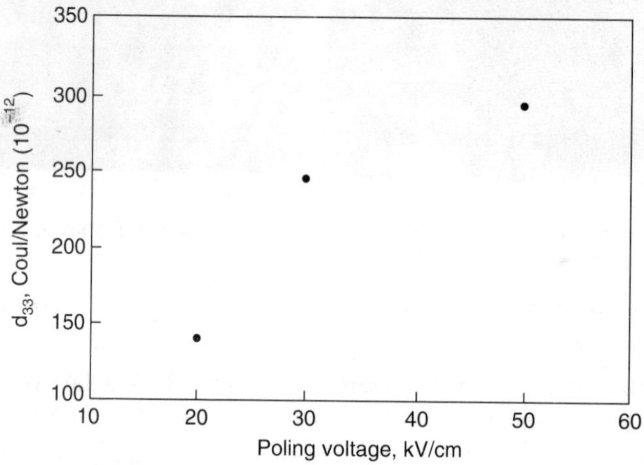

Fig. 5.6 Effect of poling voltage on piezoelectric coupling factor, d_{33} of PZT

Aqueous-sol Derived thin Films

Growth of thin films, using physical preparation methods such as, evaporation, r.f. sputtering and laser ablation has difficulty in achieving large area depositions. Moreover, it requires costly equipment. Sol-gel processing, which is a solution based method, is attractive due to easier fabrication of large area thin films at low cost. However, commonly used alkoxide precursors for producing films by sol-gel method, are highly reactive and require careful control of the hydrolysis-condensation reaction. The compatibility between the starting reagents constitutes the main problem. In addition, methoxyethanol used as a solvent, is a hazardous material. The problem has been solved in aqueous sol route to grow thin films. In this technique, nanoparticles are suspended in aqueous medium using suitable electrolyte in desired quantity. It helps to set-up an electrical double layer between the particle and the medium and prevents coagulation. The stable suspension (sol) is then deposited on substrate by spin or dip coating followed by annealing. Ferroelectric $PbTiO_3$ and antiferroelectric $PbZrO_3$ thin films have been successfully grown by novel aqueous sol route. The quality of the films obtained by using this simple technique is comparable to that obtained by other techniques. In short, the capacity of the nanoparticles to produce continuous and smooth coating on effectively large surface has been successfully used in realization of thin films. The technique could be extended to obtain large area thin films of a variety of materials, provided stable suspension (sol) in the required chemical composition could be prepared. The possibility to avoid an exposure to hazardous chemical reagent is an added advantage of the process.

■ CONCLUSIONS

In summary, the results presented in the paper obtained by using nanoprecursors reveal the technological advantages in various ways. Offering clean separation of *n*-TPP and its derivatives and enhancement in catalytic efficiency in hydrogen-oxygen combination reaction are clear examples of enhanced surface activity. Synthesis of $YBa_2Cu_4O_8$ under the condition of normal oxygen pressure is an indication of increased chemical reactivity. Lowering of sintering temperature in PZT supports enhancement in densification rate with reduction in precursor size. Growth of continuous thin films of $PbTiO_3$ and $PbZrO_3$ using spin coating is the best illustration of surface coating efficiency.

6

New Synthesis Method of Carbon and Boron Nitride Nanotubes: Mechano-Thermal Process

Nanotubular materials including carbon and boron nitride (BN) have excellent mechanical and physical properties compared with their bulk materials. The seamless graphene cylinders with a high length to diameter ratio (more than 1000) make them as superstrong fibers to enforce composite materials. Its special electronic properties varying with diameter and helicity have been used to make nanosized transistors. The smallest cylinders can store huge amount of hydrogen gas due to the large surface area and microporous structure and thus create a bright future for hydrogen-powdered cars. All these promising industrial applications require obviously a large quantity of nanotube materials in the order of kilograms or above. However, mass production in the order of kg of high quality nanotubes is still difficult. Discovery of new synthesis methods is still crucial for any application in industrial scale. A new mechano-thermal process was discovered by the current author in 1998. This mechano-thermal process consists of mechanical milling and thermal annealing processes, and can produce nanotube materials in the order of kilograms. In addition, because both milling and heating processes are fully controlled and at relatively low temperatures (less than 1400 °C), it provides an excellent opportunity to investigate formation mechanisms of nanotubes, which are not yet fully understood. Another advantage of this method is that boron nitride nanotubes can be produced readily. Boron nitride nanotubes have similar tubular structure as carbon ones, but with a better thermal conductivity, more stable electronic properties and higher chemical stability. However, BN nanotubes were produced with difficulty previously. This paper demonstrates that carbon nanotubes have been produced in graphite samples after pre-milling and subsequent annealing treatments. BN nanotubes can be prepared via either ball-milling induced nitridation reaction between pure elemental B powder and ammonia gas, or mechanically activated solid-state crystal growth. Finally, new formation mechanisms are discussed.

■ EXPERIMENTAL PROCEDURE

High-purity crystalline graphite, boron (B) and boron nitride (BN) powders were used as the starting materials. Ball milling experiments were performed at room temperature in a vertical planetary ball mill using hardened steel balls with a diameter of 25.4 mm and a stainless steel container. A schematic illustration of ball milling actions is shown in Fig. 6.1. The milling container was loaded with several grams of starting powder together with four milling balls. A starting pressure of about 300 kPa of ammonia or argon gas was established before each milling experiment. Heat treatments were conducted under a N_2 gas flow at temperature up to 1400 °C in a tube furnace. Milled and annealed samples were characterized using various analyzing techniques. Crystalline structure of samples was investigated by means of X-ray diffraction analysis (XRD) using Co radiation (l = 0.1789 nm) at room temperature. A Hitachi S4500 field-emission scanning electron microscope (FESEM) is used for imaging nanosized fibers and tubular structures were observed using a Philips EM430 (300kV) transmission electron microscope (TEM). The average metal concentrations were examined by using X-ray energy dispersive spectroscopy (XEDS) in a JEOL. (JSM6400) scanning electron microscope.

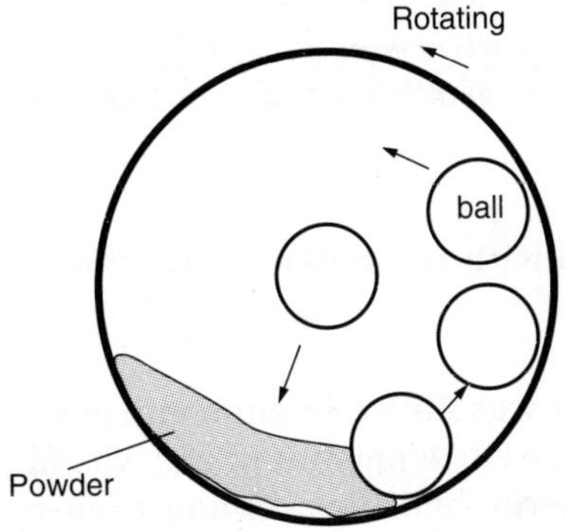

Fig. 6.1 Schematic representation of milling action

RESULTS AND DISCUSSIONS

Carbon Nanotubes

A graphite sample was first milled for 150 h and then heated at 1400 °C for 6 h in flowing nitrogen gas. Carbon nanotubes in the web form were found in the annealed sample as shown by a typical TEM micrograph in Fig. 6.2. The outer diameter of the tubes is less than 20 nm. The inner diameter is in the range of a few nanometers. No metal particle is observed inside these tubes but dark metal particles are observed around. These particles are iron contaminants from steel milling balls and the container. Multi-walled nanotubes containing a particle at tip were also observed. In these cases, the iron particles should act as catalysts for the nanotube formation, as suggested in other processes. Most nanotubes form sub-micron sized clusters. No nanotubes were found in the sample directly after milling treatment without subsequent heat treatment. Also, direct annealing of the starting graphite powder without pre-milling treatment does not produce any nanotube. Therefore, the nanotubes should grow only during annealing process from the pre-milled sample with or without the help from the iron particles. It is found that the milled sample has a nanoporous and disordered structure. Because the temperature at which nanotubes are produced is much lower than the melting point of graphite, vapour phase should not be produced. Nanotubes are produced in this special process in a solid state and thus with a new formation mechanism.

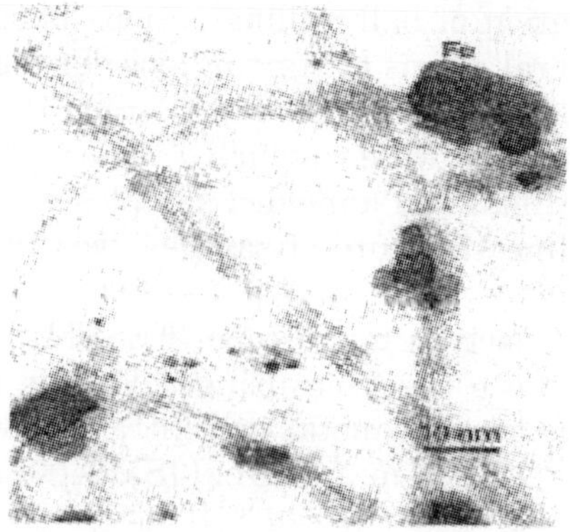

Fig. 6.2 TEM micrograph of carbon nanotubes produced by mechano-thermal process. The particles with dark contrast are iron

Boron Nitride Nanotubes

BN nanotubes can be produced with two different reaction processes.

Reactive ball milling

The starting pure boron powder was milled in ammonia gas at a starting pressure of 300 kPa for up to 150 h. During milling, a huge amount of gas was absorbed onto the fresh surface of the powder created by high-energy ball impacts and even nitridation reaction was induced during extended milling, as indicated by pressure decrease and re-increase inside the milling chamber. Nanocrystalline BN was produced at the end of milling, which serve as seeds for nanotube growth during subsequent heating process.

Subsequent heat treatment of the as-milled sample under nitrogen gas flow induced progress of the nitridation reaction at a temperature of 1300 °C for 6 h. XRD analysis reveals the formation of hexagonal BN phase and small amount of iron boride due to iron contamination from milling media (balls and container). Fine filaments extruding from BN aggregates are observed with FESEM, as shown in Fig. 6.3. TEM reveals that these filaments are hollow with a tubular, multi-walled structure and a typical high resolution image is shown in Fig. 6.4.

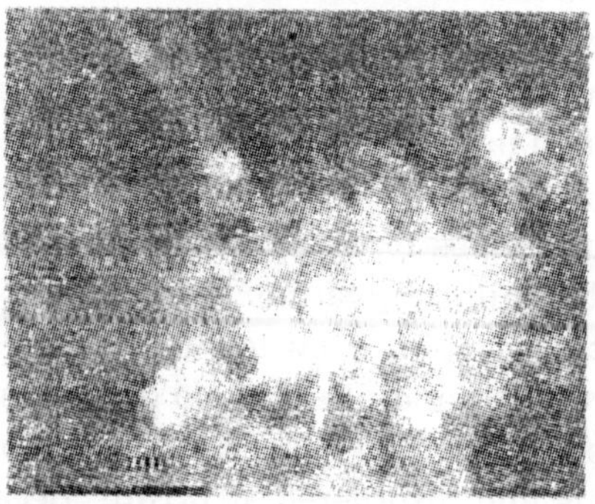

Fig. 6.3 SEM micrograph of BN nanotubes growing from BN particles

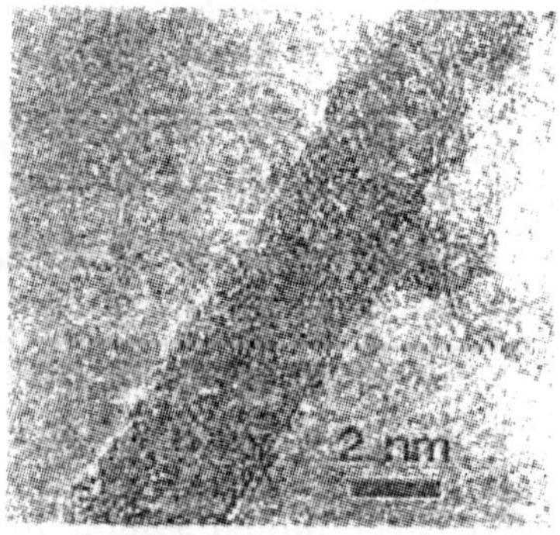

Fig. 6.4 High-resolution TEM image of BN multi-walled nanotubes

Mechanically activated crystal growth

The starting material in this case is crystalline boron nitride compound. The BN powder was milled for 140 h in nitrogen gas and then heated at 1200 °C for 10 h. Multi-walled BN nanotubes with an outer diameter of about 11 nm and an inner diameter of 3 nm were produced. Special bamboo-type tubes were observed in the same milled sample after annealing at a higher temperature of 1300 °C for 10 h (Fig. 6.5), revealing a typical catalytic role of metal particles. The BN nanotube formation process in this case is similar to carbon nanotubes because of absence of any nitridation reactions. Therefore, nanotube formation should be the result of nanocrystal growth at low temperature. Detailed mechanism is published elsewhere. This result confirms that the nanosized BN particles produced by reactive ball milling (the above case) could serve as seeds.

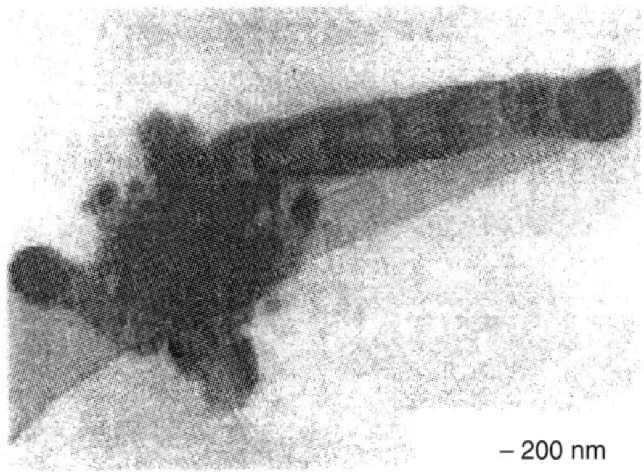

Fig. 6.5 TEM image of BN tubes with a bamboo-type structure and containing iron particles at tips

Formation Mechanisms

Nucleation

Mechanical treatment (high-energy ball milling) has an essential role in the mechano-thermal process. Without milling treatment, direct heat treatment of graphite or BN samples cannot produce any tubular structures. Nanotubes were not observed in any milled samples. These observations suggest, that, the milled materials may contain a special nucleation structure. Both XRD and electron microscopy analyses reveal that milled samples have nanocrystalline phases and highly disordered (or close to amorphous) structure. The nanocrystals are seeds and the amorphous phases are mobile atom source for tube growth. Therefore, mechanical treatment actually produces a nanotube precursor.

Nanotube growth

Nanotubes start to grow out from the precursor powders when the milled samples are heated at certain temperatures. This growth process appears to occur with a new diffusion mechanism. First, the growth temperature (close to the annealing temperature) is significantly lower than the melting points of graphite or BN, the nanotubes grow in solid state or similar to crystalization growth. At these low temperatures, surface diffusion should dominate the atomic diffusion and is the main mechanism of nanotube growth.

Metal catalysts

It was directly observed that some metal particles are involved in formation of certain nanotubes. For example, some nanotubes contain metal particles inside or at tips. These particles are known to serve as seeds and also maintain the tube open during growth. In this case, the metal particles coming from the milling media as contamination catalyze nanotube growth and therefore are useful rather than harmful to the formation process.

■ CONCLUSION

Carbon nanotubes have been produced in a graphite sample after pre-milling and subsequent annealing treatments. BN nanotubes can be prepared via either ball-milling induced nitridation reaction between pure elemental B powder and ammonia gas or mechanically activated solid-state crystal growth during subsequent thermal annealing. Metal particles coming from the milling media contamination catalyze nanotube growth and therefore are useful rather than harmful to the production process.

Surface diffusion should dominate the atomic diffusion and is the main mechanism of nanotubes growth in solid state at such low temperatures.

7

Nanostructure Formation by Mechanical Activation

Nanocrystalline materials (NCMs) are solids composed of crystallites with size less than 100 nm in at least one dimension. The NCMs can be zero (clusters), one (lamellar), two (filamentary) or three (equiaxed particles)-dimensional. Present high visibility and definition of the field can be attributed to the pioneering work of Gleiter and co-workers in 1980's. These are the materials with a large volume fraction of atoms at the grain boundaries, which can be as high as ~ 50% for 5 nm sized grains, while it is a mere 3% for 100 nm sized grains (Fig. 7.1). A number of physical and mechanical properties of materials are significantly altered in the nanocrystalline state (Table 7.1). For example, recent work has shown considerable improvement in the mechanical properties of nanostructured TiN-Al_2O_3 composite synthesized *in situ*. Over the past two decades, various facets of these materials have been reviewed in a large number of publications.

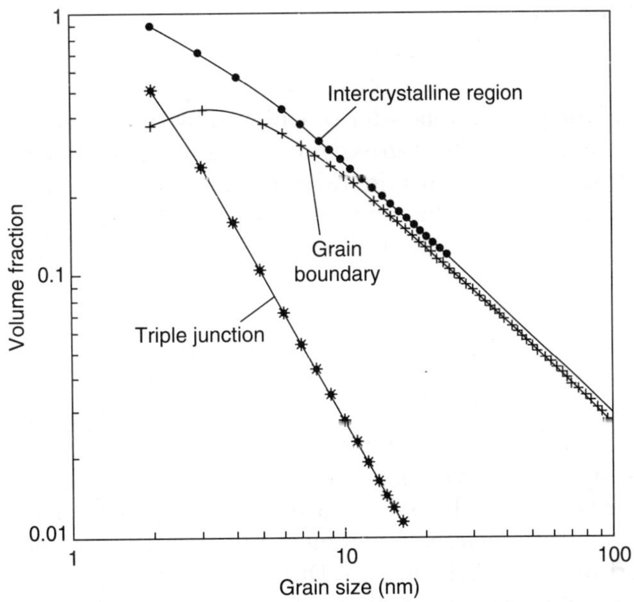

Fig. 7.1 Estimated volume fraction of grain boundaries, intercrystalline regions and triple junctions assuming grain boundary thickness of 1 nm

NCM can be synthesized starting from vapour (inert gas condensation, sputtering, plasma processing, and vapour deposition), liquid (electrodeposition, rapid solidification) and solid state (high energy ball milling, sliding wear, spark erosion, crystallization of amorphous matrix). Among the solid-state techniques, high-energy ball milling has become popular due to its simplicity, flexibility, higher productivity, scalability, and low capital investment. NCM can be produce by high-energy ball milling of individual elements (mechanical milling (MM)) or elemental blends to form alloys and intermetallics [mechanical alloying (MA)]. The formation of NCMs during MA was first suggested by Koch *et al.* and was validated by Fecht *et al.* Nanocrystalline metals, alloys, intermetallics, ceramics and nanocomposites have been synthesized in a number of systems by this technique.

Table 7.1. Changes in the properties in the nanocrystalline state

Property	Change in the nanocrystalline state
Electrical	Higher conductivity in ceramics and magnetic nanocomposites
	Higher resistivity in metals
Magnetic	Increase in coercivity until a critical size
	Decrease in coercivity below a critical size leading to superparamagnetic behaviour
Mechanical	Increase in strength and hardness in metals and alloys
	Enhanced ductility, toughness and formability in ceramics and intermetallics
Optical	Blue shift of optical spectra in quantum-confined crystallites
	Increase in luminescent efficiency of semiconductors

HIGH ENERGY BALL MILLING

High Energy Ball Mills

The energy transfer during ball milling takes place by shear or impact of the high velocity balls with the powder. An attritor was the first high-energy ball mill used for MA by Benjamin, where the movement of the balls and powder is achieved by the horizontally rotating impellers attached to a vertical shaft and set progressively at right angles to each other. The tumbler mills, which are traditionally used in mineral processing, can also be used for MA, if their diameters are sufficiently large (of the order of meters) and if mills are operated close to the critical speed beyond which the balls stick to the inner walls of the mill. For large-scale production, tumbler mills are more economical when compared to the attritor and other high-energy ball mills.

The laboratory mills, e.g., Spex 8000, Fritsch Pulverisette P-5 and uni-ball mills, though have smaller capacities, offer a higher velocity for the balls and have found wide use. In Spex mill, the vial containing the balls and powder is vibrated in three mutually perpendicular directions while in the Fritsch planetary mill (P5 and P7), the disc and the vial mounted on the disc rotate in the opposite directions, which gives a high centrifugal force to the balls. Even though the linear velocity of the balls in a standard planetary mill is higher than that of shaker mill, the frequency of impact is much more in the later case, which makes it a higher energy mill in comparison to the Fritsch mill. Due to the low milling energy in the attritor, it takes a longer period to achieve alloying when compared to the laboratory mills. It was shown that alloying in the case of Ti-Mg system takes place after 16 h in the Spex mill, while the attritor took 100 h to achieve the same result.

The uni-ball mill involves a controlled movement of grinding balls that is confined to the vertical plane by the cell walls and adjustable external magnetic field. The mill can run in high-energy (impact) or in low energy (shear) mode. In addition to the above mills, several other designs such as Fritsch pulverisette P0(single large ball in a vibrating frame mill), rod mill, modified rod mill and other specially designed mills have been used for MA/MM. In a recent advance, a modified, electric discharge assisted mechanical milling technique has been developed, where materials react in a gas atmosphere under an electric discharge. Application of low current high voltage electric impulses results in faster reactions and new synthesis and processing routes. Glow (cold) discharge milling was found to promote solid-gas reaction whereas spark (hot) discharge milling promotes fast fracturing, recrystallization, mineral reduction and solid-solid reactions.

Milling Parameters

The final attainable grain size of nanomaterials by high-energy ball mills and conventional low energy ball mills could be similar. However, the kinetics of phase formation appears to depend on the energy transferred to the powder ingredients from the balls during milling. The energy transfer is governed by various parameters e.g., the design of the mill, milling speed, type, size and size distribution of the balls, ball-to-powder weight ratio (BPR), extent of filling of the vial, temperature of milling, grinding media, milling atmosphere and the duration of milling. Joardar *et al.* have shown that nanocrystalline NiAl formation and its disordering tendency during MA are enhanced with the increase in milling intensity. Similarly, the rate of formation of nanocrystalline TiC during MA increases exponentially with the density of the balls. The size and size distribution as well as the number of balls should be optimized in order to achieve efficient packing of the vial. Dense packing of the balls decreases the mean free path of the ball, while loose packing minimizes the collision frequency.

Contamination

Majority of the MA work makes use of stainless steel (SS) or hardened chrome steel (CS) grinding materials. However, these can introduce Fe-contamination into the milled powder due to continuous wear caused by the impact and friction involved in the processes. It has been shown that Fe-contamination may reach about 18at.% during the synthesis of NiAl from Ni and Al blend using SS grinding media. However, with CS milling media, which has a better wear resistance than SS, the contamination level drops to ~5at.%. High Fe-contamination was also reported during cryomilling of Al-Mg-AlN alloy in attritor made of 316 L steel, possibly due to brittle behaviour of steel balls in liquid nitrogen. The level of contamination not only depends on the milling media but also on other milling conditions such as the type of mill, milling speed, mechanical properties of powder being milled and its chemical affinity for the milling media. For example, milling of elemental Ni in Spex mill leads to about 13at.%. Fe-contamination to attain minimum grain size while for elemental Cu it is ~ 1at.%. Contamination as large as 60at.% Fe was reported during milling of W-5Ni alloy for 20 h in a Spex mill. Contamination problem tends to increase for high-energy ball mills. In order to minimize contamination, low energy milling modes or less energetic mills such as the vibratory mill may be used. Contamination can also be avoided by using a milling media made-up of the same material as the powders.

Contamination can cause profound influence on the MA process as well as the product quality. For example, MA of elemental blend of Ni and Al has been found to trigger a highly disordered (long range order parameter, $S < 0.3$) nanocrystalline NiAl formation when milled with SS media, while partially ordered ($S = 0.55$) NiAl forms when milled in tungsten carbide.

Milling with CS media has resulted in NiAl having an intermediate value of S (0.48). The difference in the ordering characteristics has been attributed to Fe contamination since Fe is known to reduce the ordering energy of NiAl. Similarly, complete alloying and the formation of solid solution was observed when an elemental blend of $Cu_{50}Ni_{50}$ was milled in Fritsch Pulverisette (P-5) at 300 rpm for 20 h in SS and CS milling media, while no alloying was observed when milled in WC under identical conditions. Interestingly, in the case of CS and SS milling media, the Cu and Ni crystallites reached nanocrystalline state (<20 nm) within 10 h of milling, while in the case of WC media they remained coarse (>100 nm) even after 20 h of milling. This could be attributed to the change in the deformation characteristics of Cu and Ni due to Fe contamination during milling in SS and CS media. Pabi *et al.* have shown that nanocrystallization is a prerequisite for alloying during high-energy ball milling, at least in slow diffusing systems such as Cu-Ni.

Milling Temperature

It has been established that the kinetics of nanocrystal formation during milling, e.g., in TiNi, is a function of the milling temperature. Shen and Koch reported smaller crystallite size for Cu and Ni when milled at 188 K as compared to those milled at room temperature. The crystallite size of CoZr intermetallic was also smaller at low milling temperature. Milling at subzero temperatures prevent excessive welding, while fracturing is favoured due to the change in the deformation characteristics of the powder at low temperatures. Thus, estimation of milling temperature is of considerable importance. However, due to the nature of the equipment, direct measurement has practical difficulties.

There have been a number of reports on the measurement of macroscopic temperature of the mill. Davis and McDermott have measured the maximum temperature of the Spex mill to be 323 K (313 K without balls). Kimura and Kimura have reported a maximum macroscopic temperature of 445 K in the attritor. Borzov and Kaputkin have measured a temperature in the range of 373-488 K for attritor using SiC and diamond sensors. Substantial temperature rise (120 K) is also reported by Kuhn *et al.*, in vibratory mills. Thus, the macroscopic temperature during milling appears to be low and sensitive to the mill design.

Based on assumption, that the particles trapped between colliding balls (head-on collision assumed) deform by localized shear, Schwarz and Koch estimated a microscopic temperature rise of about 40 K for $Ni_{32}Ti_{68}$ and $Ni_{45}Nb_{55}$ powders in a Spex mill. Davis and Koch calculated the ball velocities and used the expression and materials of Schwarz and Koch to predict a rise in temperature ΔT of ≤ 112 K. They have also estimated the maximum ΔT to be about 350 K. Maurice and Courtney have developed an expression for the adiabatic temperature rise during milling by modeling the deformation of powder as equivalent to microforging. However, they assumed ΔT to be low such that the material is in the cold working regime. Magini *et al.* have calculated the adiabatic temperature in a planetary mill and reported the maximum microscopic temperature to be of the order of 400 K after calculating the amount of powder trapped between the balls assuming a Hertzian collision. Bhattacharya and Arzt calculating the contact temperature of the powder compact surfaces, assuming Hertzian elastic collisions, which are higher (623 K) than those calculated by others. However, they show a rapid decrease of the temperature to ambient temperature at the center of the compact. Davis and Koch calculated the ΔT due to sliding friction to be at very low (≤ 10 K). Miller *et al.* used microsecond time resolved radiometry to observe a temperature rise during impact on various materials. They observed ΔT of the order of about 400 K. Based on simple calorimetric technique, ball temperature during MA in AGO-2 planetary ball mill was recently estimated to be ~893 K in cases where compact layer of milled powder are found on the ball/vial surface.

Observation of deformation bands and slip lines by optical microscopy and the high dislocation densities and deformation bands seen in transmission electron microscopy indicate that the microscopic temperature are much below the recrystallization temperature. However, in some cases it is reported that the temperature is above the crystallization temperature of the amorphous phase. Davis and Koch studied the tempering of martensite in Fe-1.2%C steel during ball milling and inferred that the maximum microscopic temperature is about 548 K. They have also milled Bi powder and concluded that the temperature of milling is below the melting point of Bi (544 K).

■ NANOCRYSTALLINE PHASES BY MA/MM

Nano-metals and Alloys

The minimum crystallite size of metals obtained by MM seems to vary with the melting point of the element. It was argued that in case of element with low melting points, the tendency for cold welding dominates, resulting in larger crystallite sizes. Koch has compiled the minimum crystallite sizes for various elements by MM (Fig. 7.2). The figure depicts that for *fcc* elements having lower melting points (Al, Ag, Cu and Ni) the minimum crystallite size varies inversely with their melting point. However, in case of *bcc, hcp* elements and *fcc* elements with higher melting points (≥ melting points of Pd), the minimum crystallite size is virtually insensitive to their melting points. The fact that similar crystallite sizes were obtained both in conventional ball mills as well as in the high energy ball mills suggest that it is the total strain, rather than the milling energy, that decides the minimum attainable grain size by MM.

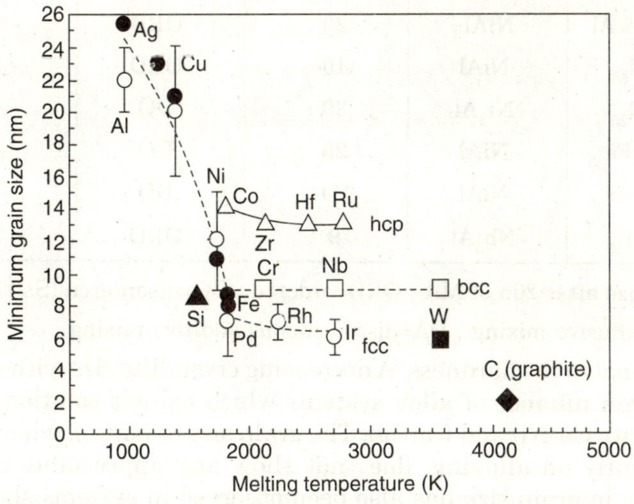

Fig. 7.2 Influence of melting temperatures on the minimum attainable crystallite size for various elements

A hitherto unknown allotropic transformation in metals like Ni and Nb during high energy ball milling of $Ni_{90}Si_{10}$ and $Nb_{80}Al_{20}$, respectively, were reported in some recent work. It was observed that metastable *fcc* to *hcp* transformation occurs in Ni, which is accompanied by a volume expansion of ~8.6% and a critical crystallite size of ~10 nm, while *bcc* to *fcc* transition takes place in Nb, which is associated with a gradual decrease in the lattice parameter of *bcc*-Nb. Positron annihilation studies evidenced a remarkable reordering of the atoms on the grain boundary and intercrystalline regions and such reordering was attributed to mechanical instability with nearly collapsing rigidity of the grain boundaries and a characteristic lattice expansion within the grains below a critical crystallite size of ~10nm. It is worthwhile to note that such critical size of structural instability is known to occur in several materials as enlisted in Table 7.2.

According to Fecht, the development of nanocrystalline microstructure during high-energy ball milling, consists of three stages. In the first stage, deformation localization occurs in shear bands containing high dislocation density. In the second stage, dislocation annihilation/recombination/rearrangement occurs to form cell/subgrain structure in nanoscale and further milling extends this structure throughout the sample. In the third stage, the orientation of subgrains becomes random. It has been suggested that the ultimate grain size achievable is determined by the minimum grain size that can sustain dislocation pile-up within the grain and the rate of recovery during milling. This means that the minimum crystallite size should

Table 7.2. Alloying mechanism in different system during MA [157]

System	*Composition*	*Product phase*	*CS (nm)*	*Product nature*	*ΔH_f (kJ/mol)*	*Alloying mechanism*
Cu-Ni	$Cu_{50}Ni_{50}$	Cu(Ni)	20	DO	2	CD
Cu-Zn	Cu-15 to 50at%Zn	α	20	DO	– 9	CD
Cu-Al	$Cu_{80}Al_{20}$	Cu(Al)	10	DO	– 8	CD
Fe-Al	Fe-25 to 50at%Al	FeAl	15	DO	– 31	CD
Ni-Al	Ni-21at%Al	$NiAl_3$	20	ORD	– 39	DA
	$Ni_{50}Al_{50}$	NiAl	10	ORD	– 72	DA
	$Ni_{75}Al_{25}$	Ni_3Al	30	DO	– 42	CD
NI-Al-	$Ni_{40}Al_{40}Fe_{20}$	NiAl	25	DO	—	CD
Fe/Cr	$Ni_{40}Al_{40}Cr_{20}$	NiAl	20	DO	—	CD
Nb-Al	$Nb_{25}Al_{75}$	Nb_3Al	9	ORD	– 20	CD

CS=crystallite size after 20h of MA ; ORD=ordered ; DO=disordered (S≤03);

CD=continous diffusive mixing ; DA=discontinuous additive mixing.

be inversely proportional to the hardness. A decreasing crystallite size with solute concentration has been observed in a number of alloy systems which exhibit solution hardening such as Cu(Fe), Ti (Cu), Nb(Cu), Cu(Ni) and Cu(Co). The grain size of Ni(Co), where the hardness does not change significantly on alloying, does not show any appreciable change during MA. Similarly, an increase in grain size has also been observed in systems showing solid solution softening such as Ni (Cu), Fe(Cu) and Cr (Cu).

Nano-Intermetallics

Over the past one decade, a considerable volume of work has been reported on the synthesis of a large number of nano-intermetallic compounds through MA. The bulk of these efforts have been on the aluminides and silicides. It is belived that the nanostructured intermetallics synthesized by MA has the potential to induce formability in these otherwise brittle materials. Fig. 7.3 shows that the increase in hardness is quite significant (by 4 to 5 folds) in case of metals on nanocrystallization, while in case of intermetallics (Nb_3Sn) it is marginal (10-20%).

Dymek *et al.* have demonstrated sufficient compressive ductility (>11.5%) in fine grained (<1μm) NiAl. Jain and Christman have shown that nanocrystalline FeAl (Fe-28Al-2Cr crystallites of 80 nm) is brittle in tension, but is superplastic in compression.

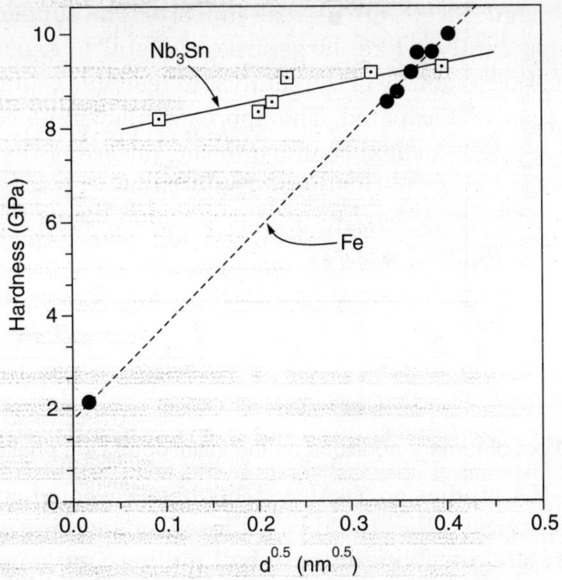

Fig. 7.3 Variation in hardness with average grain diameter (d) for Fe and Nb_3Sn

Aluminides

Among all aluminides, MA of Ni, Ti and Fe aluminides have so far received major attention apparently due to their potential application in aerospace and automotive industries. MA has also demonstrated its ability in the synthesis of several other aluminides including ternary aluminides e.g., Cu-Al, Nb-Al, Mo-Al, Zr-Al, Al-Ni-Fe, etc.

Ni-Aluminides—The formation of nanocrystalline Ni-aluminides, such as $NiAl_3$, NiAl and Ni_3Al through MA has been established at various compositions in the binary Ni_xAl_{100-x} elemental blends. Though a considerable extension of phase field in the Ni_3Al, NiAl and even in the line compound $NiAl_3$ have been observed under intensive milling, Ni_2Al_3 and Ni_5Al_3 phases were found metastable under similar conditions. In a marked contrast to the gradual temperature rise (~348 K) during MA of Ni-aluminides in laboratory ball mill under air, the formation of NiAl in Spex 8000 mill under Ar atmosphere was accompanied by an exothermic reaction within a short duration of milling following an interruption after 2 h of continuous milling. A similar observation has also been reported recently on low energy ball milling of $Ni_{50}Al_{50}$. Recent studies by Liu *et al.* have revealed a delayed initiation of the exothermic NiAl phase formation reaction on addition of ternary alloying element such as Ti and Fe as indicated in Fig. 7.4. The occurrence of explosive reaction was also observed on opening the vial following milling. This suggests the possible role of energy liberated during oxidation of Al in promoting the Ni-Al reaction. On the other hand, the absence of catastrophic reaction when milled in air could be attributed to the continuous oxidation of the elements and the slow diffusion between the oxide-coated ingredients, which result in reduced reaction kinetics. Pabi and Murty have observed wide phase fields for the Ni-Al system in the nanocrystalline state (Fig. 7.5), which was explained by thermodynamic considerations.

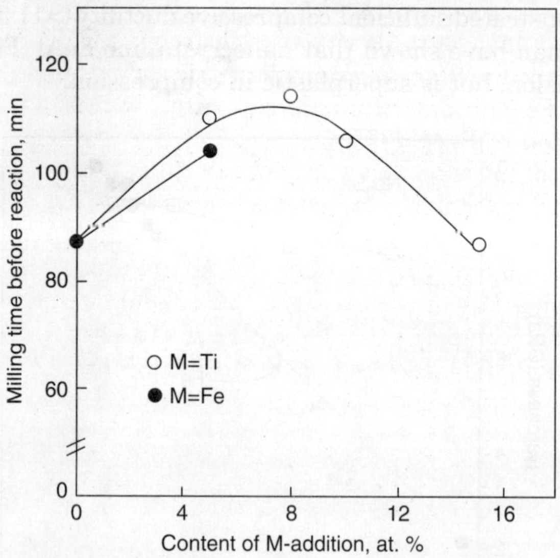

Fig. 4 Effect of ternary additions on the initiation of NiAl phase formation

Ti-Aluminides—MA studies on Ti-Al systems have revealed stiff resistance to alloying. The formation of Ti-Al intermetallics could be achieved on adopting a two-stage 'mechanically activated annealing process' involving intense ball milling and subsequent annealing. Some reports on direct synthesis of Ti-Al intermetallics by MA are also available. Investigation by Park *et al.* have shown successful synthesis of TiAl and Ti_2Al through MA in vibratory mill with stainless steel grinding media and 12.7 and 19.1 mm diameter grinding balls. However, the extent of Fe-contamination and its possible role in triggering such a reaction during MA was not studied.

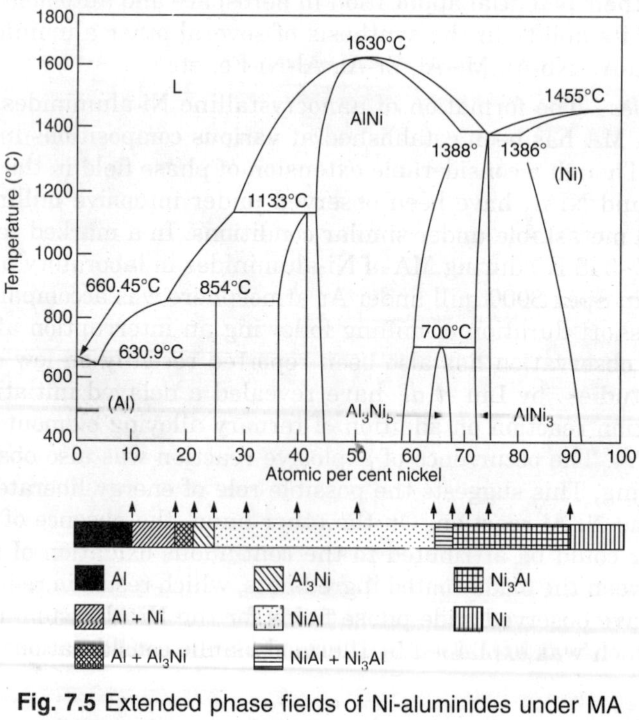

Fig. 7.5 Extended phase fields of Ni-aluminides under MA

Fe-Aluminides—MA of $Fe_{20}Al_{80}$ and $Fe_{25}Al_{75}$ blends in planetary ball mill has shown $FeAl_3$ formation. The $FeAl_3$ formation took 40 h from $Fe_{25}Al_{75}$ blend as against 65 h from $Fe_{20}Al_{80}$. This is possibly due to the off-stoichiometric composition. No intermetallic compound has been detected in Fe-87.5 at.%Al blend even after 80 h of MA. Subsequent annealing at 773 K, however resulted to $FeAl_3$. The formation of $FeAl_3$ at non-stoichiometric composition may be an indication of an extended $FeAl_3$ phase field, though, any possible Al loss due to oxidation may also be responsible for such observation. MA of Fe-rich composition, e.g., Fe-25at.%Al, which lies in the Fe_3Al phase field, has been found to produce FeAl. Subsequent annealing, on the other hand, has led to B2-Fe_3Al instead of the usual DO_3 structure. The formation of FeAl during MA in preference to $AlFe_3$ could be attributed to its more negative ΔH_f (– 31.8 kJ/mol) when compared to that of $AlFe_3$ (– 18 kJ/mol). The formation of metastable Fe_2Al_5 through MA was also demonstrated within 0.5h in Spex 8000 mill, which finally transformed to the more stable FeAl after 5 h.

Silicides

Silicides of Ni, Fe, Ti and Mo have received wide interest in recent years, because of their potential application as structural materials in the field of microelectronics and electrical technology.

Ni-Silicides—The initial efforts by Redlinski and Calka on the synthesis of Ni-Si intermetallics through MA of an equiatomic blend failed even on prolonged milling for 1000 h. However, subsequent annealing at 973 K for a period of 1 h led to the formation of NiSi. On the other hand, studies on MA of $Ni_{100-x}Si_x$ (x=25, 28, 33, 40 and 50) in Fritsch P-5 at a very high vial speed of 642 rpm revealed crystalline phase formation in Ni_3Si, Ni_3Si_2 and NiSi compositions and an amorphous phase in Ni_5Si_2, Ni_2Si compositions. Recently, Datta *et al.* studied the MA characteristics of the entire range of composition in Ni-Si system. They have observed that only congruent melting silicides form in the nanocrystalline state and the formation of non-congruent silicides is suppressed even at their equilibrium compositions. Thus, in the nanocrystalline state the phase fields of silicides are given by the metastable phase diagram instead of the stable phase diagram as shown in Fig. 7.6. (*a*) and (*b*).

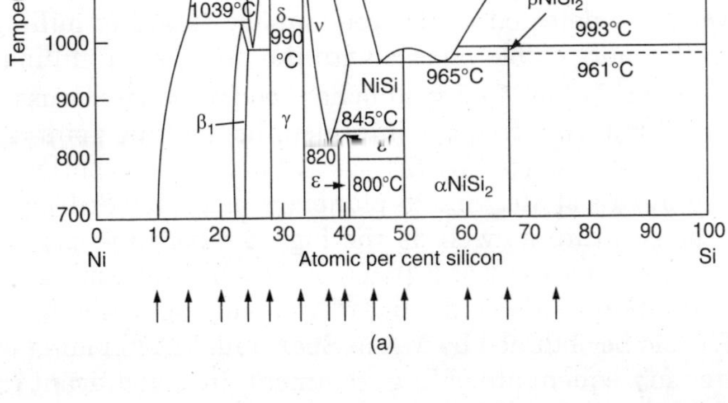

(a)

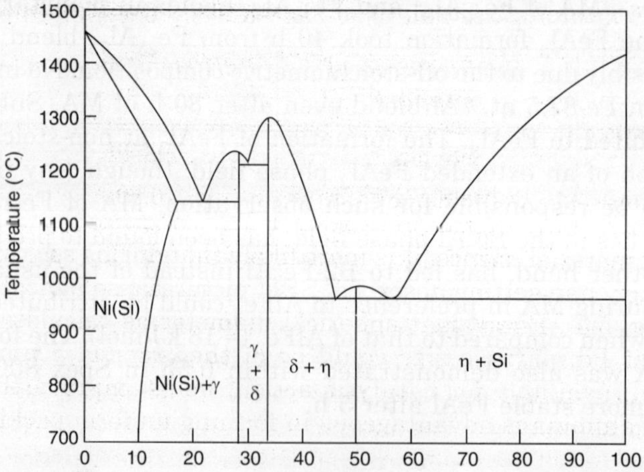

(b)

Fig. 7.6 (*a*) Equilibrium Ni-Si phase diagram indicating various blend compositions subjected to MA and (*b*) Metastable phase diagram nanocrystalline state

Fe-Silicides—MA efforts with elemental $Fe_{33}Si_{67}$ blend have shown the formation of a mixture of low temperature-tetragonal α-$FeSi_2$, high temperature-orthorhombic β-$FeSi_2$ and the cubic FeSi phases. MA studies on $Fe_{100-x}Si_x$ (x=5, 25,37.5 and 50) in Fritsch P-7 at a BPR of 12 have revealed the formation of FeSi, Fe_2Si, Fe_5Si_3 and $FeSi_2$ respectively. However, MA of Fe-6.5 wt%Si has shown the formation of a solid solution. In fact extremely high intensity MA has demonstrated the formation of Fe-Si solid solution over an extended phase filed up to Fe-37.5 wt.%Si. On the other hand MA studies in the Si-rich composition regimes in horizontal ball mill have revealed the formation of β-$FeSi_2$ phase for Si>70at.% and ε+β-$FeSi_2$ at 50at.%<Si<70at.%.

Ti-Silicides—In the initial studies on MA of Ti-Si system by Veltl *et al.* in planetary ball mill, no intermetallic formation in the intermediate stages of milling have been detected possibly due to fast amorphization reaction during high energy milling. However, the generation of TiSi intermetallic phase prior to amorphization during MA has been observed within 4 h in Vibro mill at higher BPR. MA result reported by Oehring and Bormann, and Park *et al.*, demonstrate the formation of crystalline Ti_5Si_3, which remained stable even on prolonged milling. Such a high stability of Ti_5Si_3, against amorphization during milling is in conformity with its large negative free energy of formation (~ – 71 kJ/mol at 673 K). Recent studies on MA of elemental blends of compositions $Ti_{33}Si_{67}$ and $Ti_{42}Si_{58}$ have shown the formation of crystalline Ti_5Si_3 and Ti_5Si_4 embedded in an amorphous matrix after 60 h of milling in Vibro mill. $TiSi_2$ and TiSi phases were also detected in the preliminary stages of milling (4 h) in this study, which paved way to the more stable phases with the progress of milling.

Mo Silicides—So far MA of MoSi system has been primarily focussed on the Mo-disilicide ($MoSi_2$), which is a well-known refractory intermetallic (melting point=2323 K). $MoSi_2$ by MA in Spex 8000 mill has been reported by Schwarz *et al*. More recently, nanocrystalline $MoSi_2$ has been synthesized by MA of $Mo_{33}Si_{67}$ in planetary ball mill (Fritsch P-5) at a disc speed of 280 rpm. The low temperature as well as the high temperature phases viz. α and β-$MoSi_2$ evolved during MA, with the extent of β phase increasing with milling time. Lee *et al.* have reported that the formation of $MoSi_2$ during MA is quite sensitive to the mill design. While both α and β-$MoSi_2$ could be obtained by MA in Spex mill. MA in planetary mill could not yield the silicide even after subsequent annealing treatment. In a significant report, Liu and Magini have shown that high energy ball milling results in a self-propagating high temperature synthesis (SHS) reaction leading to stable α-$MoSi_2$, while low energy milling leads to metastable β-$MoSi_2$ without any such reaction.

■ NANOCOMPOSITES

In situ synthesis of nanocomposites by MA is advantageous, as it cuts down the production steps and assists better intermixing of the constituents and leads to enhanced properties. Zhu *et al.* have synthesized Pb-Al and Fe-Cu nanocomposites by MA. In a similar attempt, Provenzano and Holtz have shown the formation of Ni-Ag and Cu-Nb nanocomposites by MA. Du *et al.* have produced Al-BN composites by high-energy ball milling. Naser *et al.* have shown that no grain growth occurs in the matrix close to its melting point when Cu and Mg are reinforced with nanocrystalline Al_2O_3 by mechanical alloying. Wu *et al.* have reported the formation of nanocrystalline TiC in an amorphous Ti-Al matrix by MA. A solid state reaction leading to the formation of AlN and AlB_2 occurred during further processing of these nanocomposites. TiAl-Ti_5Si_3 nanocomposites were reported by Liu *et al.* by thermally activated MA. They attribute the formation of these nanocomposites to the crystallization of amorphous phase obtained by MA in Ti-Al-Si system. These nanocrystalline compounds appear to be stable and no significant coarsening was observed even after heat treatment for 1h at 1273K. Similar nanocomposites have been obtained by Senkov *et al.* on MA and subsequent heat treatment of a mixture of TiH_2 and Al-Si alloy powders. $NbAl_3$-NbC nanocomposites have been obtained by MA of Al and Nb powders. NbC in this case appears to have formed by the reaction of Nb with the methanol, which was added as the process control agent. Manna *et al.* have recently shown the formation of Cu-$NbAl_3$ nanocomposite by the codeposition of nano-crystalline $NbAl_3$ during the electrodeposition of Cu Pfullmann *et al.* have mechanically alloyed TiAl base alloys with Si to synthesize nanocomposites with Ti_5Si_3 dispersions in TiAl and Ti_3Al and found significant improvement in microhardness in the nanocrystalline state (Fig. 7.7). Improved tribomechanical properties have been observed in nano WC-Co and ultrafine Ti (CN) based cermets containing nano-Ni binder. These have been attributed to better interdiffusion leading to improved microstructural homogeneity and enhanced properties.

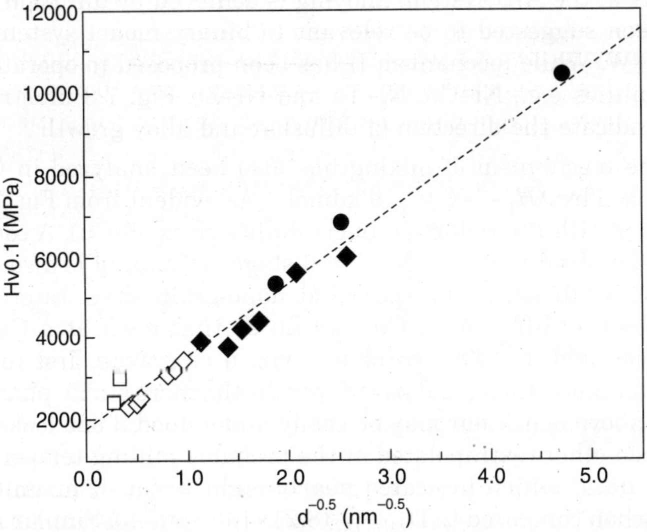

Fig. 7.7 Change in the microhardness of Ti-Al alloys with $d^{-1/2}$ (•), mechanically alloyed Ti-46at.%Al-5at.%Si and Ti-20at.%Al-17at.%Si-95at.%Nb; (■), mechanically alloyed binary Ti-Al; (□) conventional powder-processed material; (•), ingot material of similar composition

Phase Formation Mechanism During MA

The usual phase formation/transformation processes are controlled by the chemical driving force provided by the free energy/enthalpy difference. Two different conditions are encountered during MA systems with ΔH_{mix}<0 thermodynamically favours alloy formation. However, the occurrence of the reaction during MA from elemental blends is likely to be decided by the kinetic factors. The system has to cross the activation energy barrier by the mechanical energy impounded into the ultra-fine-crystal ensembles and once it crosses this barrier, the reaction is sustained by an enhanced downhill diffusion. On the other hand, systems with $\Delta H_{mix} \geq 0$ lack chemical driving force for alloying and as such, the reaction is thermodynamically not feasible. However, MA has shown several examples where alloying occurred in spite of such barrier. Several mechanisms have been suggested by different groups e.g., the role of grain boundary stored energy, oxygen induced mixing, capillary pressure, dislocations and coherent interface. It may be noted that all these mechanisms suggested for systems with positive heat of mixing could be applicable to systems showing negative enthalpy of mixing as well.

In case of nano phases produced during MA, the contribution of interfacial energy term becomes comparable with the chemical free energy and hence plays a critical role in the stability of equilibrium phases in the milled product. For instance, precipition of second phase particles in nanocrystalline supersaturated solid solution is difficult due to further increase in interfacial areas. For this reason, reactions like discontinuous precipitation or peritectoid transformations are expected to get suppressed in nanocrystalline alloys. The high dislocation density in the milled products due to heavy cold work involved in MA also contributes to the enhanced diffusivity values and significantly influences the alloying mechanism. So far, alloying in nanomaterials induced by ball milling seems to have three major types of mechanisms as discussed below:

Continuous Diffusion (CD)

This involves a continuous change in the lattice parameter of the solvent with the progress of milling at the expense of the depleting solute content. Zbiral *et al.* have proposed two different mechanisms (I and II) occurring via CD mode. In mechanism I, for a lower solubility of A in B compared to B in A in the A-B system, alloying is achieved by diffusion of the B into A. Such mechanism has been suggested to be relevant to binary model systems e.g., Ni+Al, Ni+Ti, Ni+Cr, Ni+Mo, Ni+W. While mechanism II has been proposed to operate in the systems with good mutual solubilities e.g., Ni+Cu, Ni+Ta and Ni+Zr. Fig. 7.8 illustrates the two alloying mechanisms and indicate the direction of diffusion and alloy growth.

Recently, the mechanism of mixing has also been analyzed in Cu-Zn model system, which is characterized by $\Delta H_f \sim -6$ to -9 kJmol^{-1}. As evident from Fig. 7.9*a*, the formation of α and ε-phases occur with the reduction in crystallite sizes of (Cu). A concomitant increase in the amount of Zn dissolved in Cu at the initial stages of milling is also estimated (Fig. 7.9*b*), which indicates fast diffusion of the species at nano-grain sizes. Interestingly, alloying was found to initiate by the diffusion of Cu into Zn so that even at a Cu-rich composition in equilibrium α phase field, the Zn-rich phases viz. η or ε were first to form. These Zn-rich phases were however metastable, and paved way for the more stable phase (s) at the particular compositions. The above behaviour may be easily understood if one looks into the diffusivities of Cu and Zn in each other (extrapolated at the probable milling temperature of ~473 K from high temperature data) which indicates nearly eight order of magnitude higher $D_{Cu\ in\ Zn}$ (2.99×10^{-14}cm^2/s) when compared to $D_{Zn\ in\ Cu}$ (8.21×10^{-23} cm^2/s). Similar mixing behaviour has also been observed in Cu-Ni system as reported by Pabi *et al.*

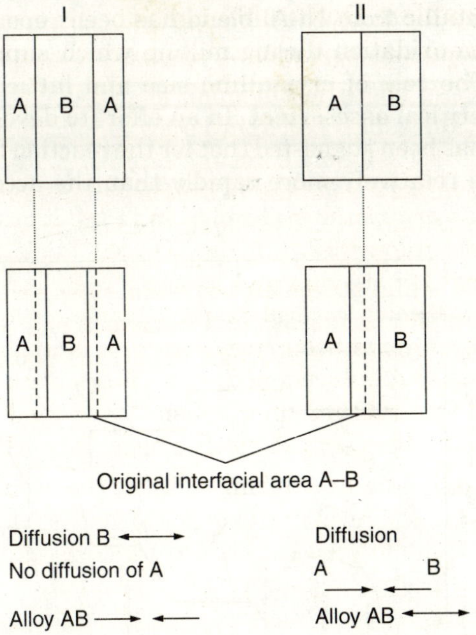

Fig. 7.8 Mechanism for continuous diffusive mode of alloying

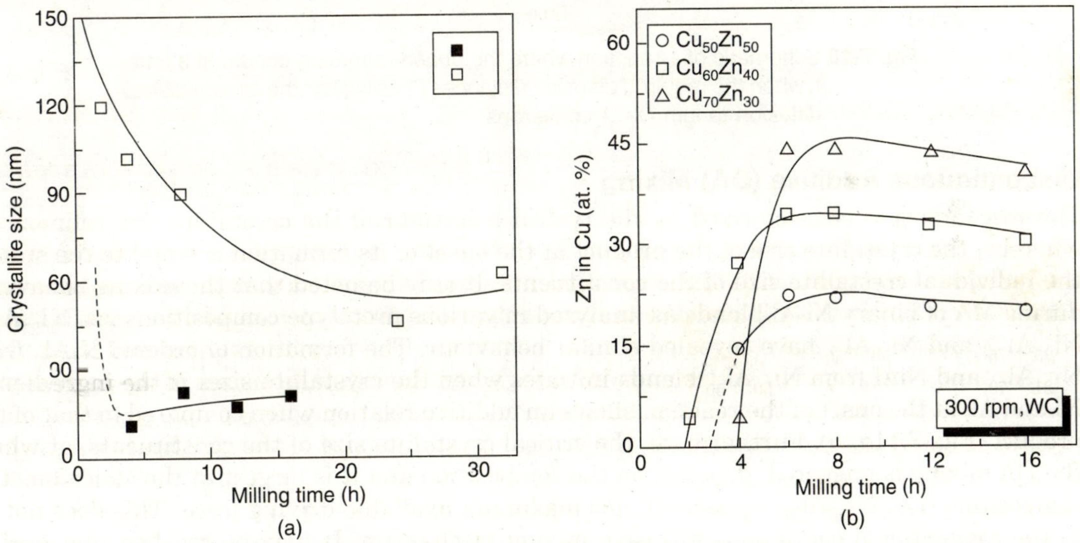

Fig. 7.9 (*a*) Changes in the crystallite size of α and ε and (*b*) variation in the solubility of Zn in Cu during MA of Cu-Zn blends

Self-sustaining Reaction (SR)

Though, in real sense SR mechanism is more a reaction synthesis mechanism rather than a MA mechanism, it deserves attention because of the indications in several reports that similar process possibly play the crucial role in the overall alloying behaviour by increasing the milling temperature drastically, which in turn promotes the desired reaction. For example, the synthesis of Ni-Al intermetallic from Ni-Al blend has been reported to be associated with an exothermic reaction due to oxidation during milling which supports the reaction leading to generation of phase(s). The role of crystalline size and lattice strain, however, has been overlooked and warrants critical assessment. In an effort to develop a well-defined criteria for self sustaining reaction it has been suggested that for the reaction to occur the required diffusion distance, l, must decrease relatively more rapidly than the accomplished diffusion distance (X_D) increases (Fig. 7.10).

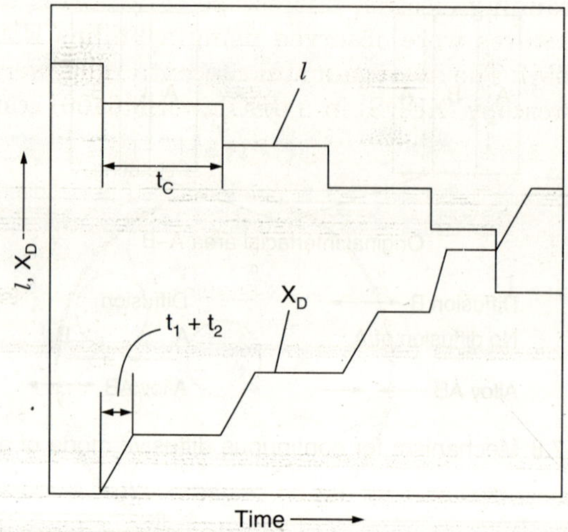

Fig. 7.10 Schematic of a reaction where the phase formation occurs at a time, t, when the required diffusion distances (l) as well as the accomplished diffusion length ($X^{\sim}_D$) converges

Discontinuous Additive (DA) Mixing

The mechanism is characterized by the typical behaviour of the crystallite size refinement, whereby, the crystallite size of the product at the onset of its formation is equal to the sum of the individual crystallite size of the constituents. It may be noted that the mixing behaviour during MA of binary Ni-Al blends as analyzed in various prototype compositions viz. $Ni_{21}Al_{79}$, $Ni_{25}Al_{75}$, and $Ni_{50}Al_{50}$ have revealed similar behaviour. The formation of ordered $NiAl_3$ from $Ni_{21}Al_{79}$ and NiAl from $Ni_{50}Al_{50}$ blends initiates when the crystallite sizes of the ingredients, Ni and Al, at the onset of the reaction shows an additive relation when compared to that of the product (Fig. 7.11*a*, *b*). Furthermore, the critical crystallite size of the constituents, at which the DA mixing is triggered, depends on the composition and it is largest at the stoichiometric composition (*i.e.*, $Ni_{50}Al_{50}$) apparently for maximum available driving force. This does not fit to the conventional continuous diffusive mixing mechanism. It may be recalled that earlier reports have indicated the association of a sudden exothermic reaction in the vial during the synthesis of NiAl from a stoichiometric blends composition, which is suggestive of reactive mixing as stated in the previous section. However, in reactive mixing the reaction front usually propagates in unpredictable direction and is, therefore, unlikely to show additive rule as observed by Pabi *et al*. In addition, as milling in these studies were performed under liquid (toluene) bath, high temperature build-up at the collision point is unlikely. The temperature estimates from earlier studies points to a modest rise in temperature (~473-573 K). It is noteworthy to mention in this connection that studies on external heating of Ni-Al powder compacts in atomic volume of 1:3 at a heating rate of 8.33×10^{-2} K/s shows the initiation of a self-propagating high temperature reaction (SHS) at ~ 823 K. While, studies on Ni-Al foils at

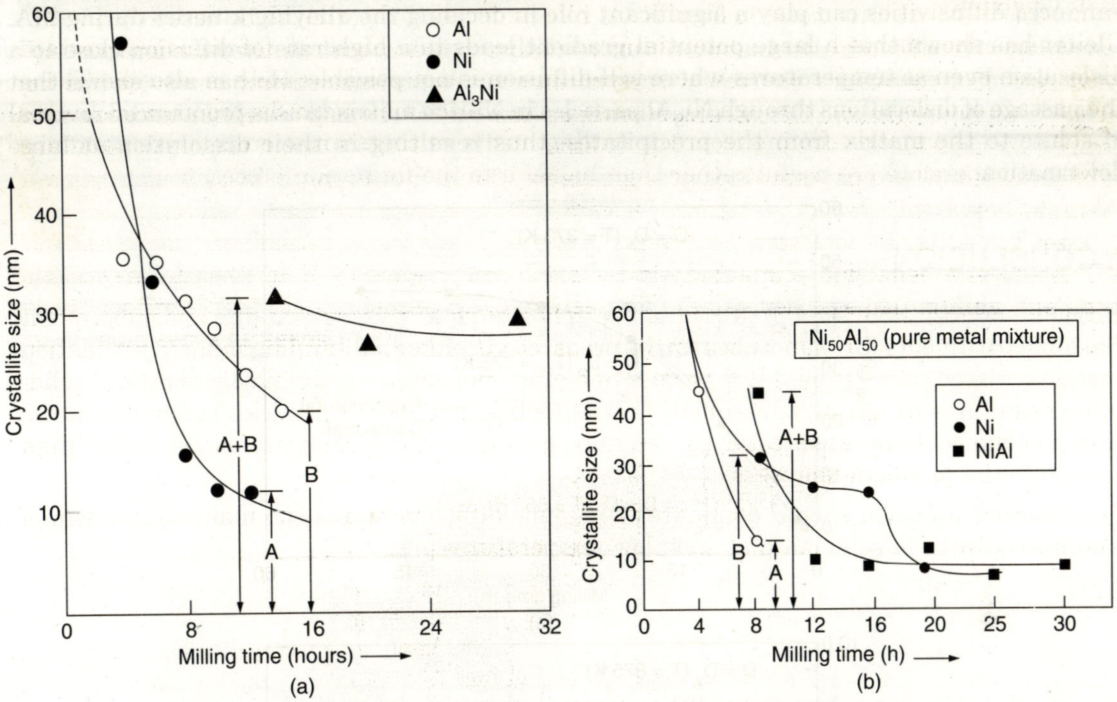

Fig. 7.11 Discontinuous additive mixing in the formation of (*a*) $NiAl_3$ from $Ni_{21}Al_{79}$ and (*b*) NiAl from $Ni_{50}Al_{50}$ blends during MA

Table 7.3. Critical grain size below which structure/property are different from conventional coarse grade materials

Material	*Critical Grain Size (nm)*	*Structure/Property*		*Ref.*
		Nano	*Conventional*	
$BaTiO_3$	120	Cubic	Tetragonal	84
$BaTiO_3$	120	Variable T_{Curie}	Constant T_{Curie}	85
Cr_2O_3	>80	Superparamagnetic	Antiferromagnetic	86
TiO_2	50	Anatase	Rutile	87
Y_2O_3	~13	Monoclinic	Cubic	88
ZrO_2	8-26	Tetragonal	Monoclinic	89
ZrO_2 (YSZ)	24-30	Reduced thermal conductivity	Regular	90

overall compositions between Ni-25 to 75at.%Al have indicated that the initiation of SHS required a temperature of greater than or equal to the melting point of Ni. Thus, the discontinuous additive mode of mixing is quite unique, which may, however, be looked upon as a low temperature version of reaction synthesis.

Various alloying mechanisms during MA operating in some of the binary and ternary metallic systems are presented in Table 7.3.

Alloying Kinetics During MA

Mechanical deformation is well known to enhance the diffusion rate and the process has been termed as "mechanical diffusion" by Balluffi and Rouff. It is envisaged that such deformation enhanced diffusivities can play a significant role in deciding the alloying kinetics during MA. Gleiter has shown that a large potential gradient leads to a high rate of diffusion close to a dislocation even at temperatures where self-diffusion is not possible. He has also shown that the passage of dislocations through Ni_3Al particles in Ni-rich matrix causes pronounced reversal of solute to the matrix from the precipitates, thus resulting in their dissolution at large deformation.

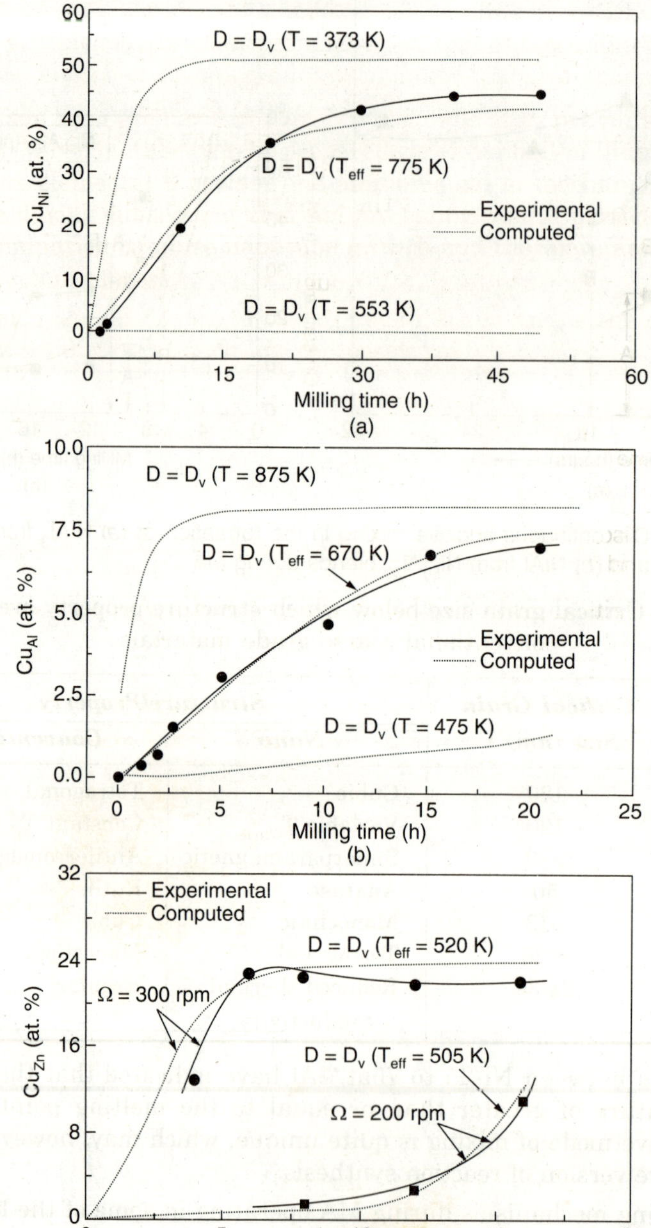

Fig. 7.12 Experimental and calculated variation of solubility of Ni, Al and Zn in Cu for $Cu_{50}Ni_{50}$, $Cu_{80}Al_{20}$ and $Cu_{70}Zn_{30}$ as a function of milling time

Recent development of a kinetic model based on a modified version (MICCM) of the iso-concentration contour migration (ICCM) method developed earlier to estimate the rate of diffusion controlled dissolution in a two phase planar and multilayered aggregate, has been a significant step in kinetic analyses of the MA process. Fig. 7.12 shows the experimental and calculated results of the respective variation of solubility of Ni, Al and Zn in Cu for $Cu_{50}Ni_{50}$, $Cu_{80}Al_{20}$ and $Cu_{70}Zn_{30}$, respectively, as a function of milling time. These experimental results are obtained from the XRD peak shifts of Cu in the above mentioned elemental blends during MA. The results clearly indicate that the effective mass transport phenomenon operative during MA attains a rate intermediate between that of volume diffusion and grain boundary diffusion. The enhanced diffusion rate in course of MA has been considered to be equivalent to that of volume diffusion at some elevated temperature T_{eff}, that has been determined by comparing the experimental values of solid solution rate with that predicted by the MICCM model. The results of these analyses demonstrate that the equation of ballistic diffusion is not able to

Table 7.4 Variation in the T_{eff} and the homologous temperature. θ_{eff} obtained from experimental data through MICCM model

Composition	*rpm*	$T_{solidus}$ *(K)*	T_{eff} *(K)*	θ_{eff} $(T_{eff}/T_{solidus})$
$Cu_{70}Zn_{30}$	200	1190	505	042
$Cu_{70}Zn_{30}$	300	1190	520	044
$Cu_{60}Zn_{40}$	300	1150	560	049
$Cu_{50}Ni_{50}$	300	1533	755	049
$Cu_{20}Ni_{80}$	300	1633	785	048
$Cu_{80}Al_{20}$	300	1317	670	051

account for the extremely large enhancement in the diffusivity values as encountered in MA. It is also clear from the results of the $Cu_{70}Zn_{30}$ that though the alloying kinetics are remarkably slower at milling disc speed of 200 rpm when compared to that at 300 rpm in the planetary ball mill P-5, the T_{eff} changes only marginally from 505 to 520 K. Furthermore, the homologous temperature ($T_{eff}/T_{solidus}$) lies in the range of 0.42 to 0.51 (Table 7.4) in each of these systems, indicating that the effective mass transfer during MA is related to the solidus temperature of the concerned system. It is also evident from these calculations that the crystallite size of the constituent(s) has to be reduced to nanoscale (<100 nm) to achieve any significant rate of alloying by high-energy ball-milling. These developments would suffice for predicting the alloying kinetics in diverse systems.

■ CONCLUSION

Some of the emerging facets of nanostructure formation by mechanical alloying/milling (MA/MM) were reviewed and the success of mechanical route in the synthesis of large number of intermetallics and solid solutions was emphasized. The studies in this field have revealed that various milling parameters, milling temperature, nature of the product(s) and the number of phases present during MA have a pronounced influence on limiting the attainable grain size product phases. Three types of alloying mechanisms were identified, viz. continuous diffusive (CD), self-sustaining reactions (SR) and the discontinuous additive (DA) mode. The role of enthalpy of formation, ΔH_p and the ordering characteristics of the phases during alloying in some of the Al and Cu based systems was emphasised. The DA mechanism was evident only in the case of NiAl and $NiAl_3$ phases having a very highly negative enthalpy (ΔH_p of –72 and –39kJ mol^{-1}, respectively) as well as an ordered structure in the as-milled state, while in rest of the phases CD mechanism was observed. A switch over from DA to CD mode occurred on disordering of the as-milled NiAl by ternary additions. The kinetic analysis of the phase formation process during MA, on the other hand, showed that the effective mass transfer rate during MA by CD mode attains a value in between the volume diffusion and grain boundary diffusion, and the rate was not so sensitive to the milling intensity. Moreover, this rate in Cu-Ni, Cu-Zn and Cu-Al systems could be correlated with the melting temperature of the systems. It was also revealed that the enhanced rate of mass transfer caused by deformation during MA is inexplicable by the exsisting ballistic diffusion model.

8

Bio-Engineering Nanotechnology

Until recent times nanotechnology was more related to materials science, electrical engineering, physics and chemistry and there has also been a considerable progress in visualization, manipulation and construction of materials at nanometer scale. Nanotechnology is presently breaking new grounds in the life science especially biomedical devices including biotechnology.

Living cells are very sensitive to environment and even particles of the size of 5 nm may induce cell to behave in an abnormal way. The reason for such behaviour arises mainly from interfacial forces and quantization effects. Therefore, when foreign materials are inserted in any living body, an unfamiliar world can cause these cells to deviate from their normal functions, making nano biotechnology as an important field of study. There are basically two important broad bands are present in the study of nano biotechnology; first one consists of producing multi-component assembly, self assembly as well as simple component particles. Apart from these surface modifications techniques aimed at nano featuring and patterning surfaces which have specific topography and specific chemistry.

Properties, types and its applications as well as potential applications with example of modifying the surface by gold particles using DNA codon sequences have been considered preliminarily. Further, the phenomenon of ion transport across cell membranes, followed by active ion transport, discussing proton pumps and its applications in drug delivery techniques also are included.

■ ASSEMBLIES

Assembly is an atomic arrangement in desired fashion with greater flexibility and precision. It as great application in construction of molecular sized tools. The modification and design of such assemblies will provide greater progress. These can be adopted for making very useful molecular machines and motors, which could power tiny pumps to open or close tiny valves. Certain prevailing biopolymers could also be useful tools; for example DNA, primarily used for encoding information could produce structures as complex as truncated octahedron.

In general there are two categories of assemblies; the self assembly and the positional assembly. In self assembly, the parts move randomly under the influence of thermal noise and possible mutual orientations are explored until a stable arrangement is obtained. For example two complimentary strands of DNA in a solution would eventually find each other and form a double helical configuration. During positional assembly certain restoring forces keep the pair positioned at a particular location and two parts are assembled when they are moved into a closed proximity and linked together. However the positional uncertainty depends on the restoring forces and temperature. Thermal noise plays a very important role in holding the various parts of the assembly. It does not provide for the positional certainty, but restricts the assembly to certain range of positions.

Self and positional assemblies are inter linked. All of the assemblies lie between these two types on a scale depending on positional uncertainty. When the positional uncertainty is high, it is more of a self assembly and in the case when positional uncertainty is high it is more of a positional assembly. The requirements of the building blocks in the above two cases are not the same. In positional assembly, the building blocks must not only link to each other, but they must also bind and release from the positional device. Secondly the linkage groups between the blocks in the positional assembly are very strong and rigid as compared to the multiple weak bonds of a self assembly. In the self assembly, these linkages provide a high degree of specificity in inter-building block reactions. In positional assembly the building blocks are more reactive, so the reactions need to be controlled. The positional assembly building blocks should not be allowed to encounter with each other in uncontrolled random orientations.

■ DNA AS A MATERIAL FOR NANOTECHNOLOGY

Existing photolithographic technique to form structures of less than 100 nm size is very expensive, therefore it was required to explore for new methods which could utilize assembly of even smaller molecular buildings to form larger devices preferably through self assembly. Biological macromolecules viz., nucleic acids and proteins may be used as components for the construction of nanostructured systems. DNA molecules with the conventional nucleotides help create specific shapes, topologies and arrangements of secondary and tertiary structures. DNA is not only a unique information carrier and occurs in a large variety of structures but also displays comparably high physio-chemical stability including mechanical rigidity. Sticky ends of DNA provide the most predictable, diverse, reliable and programmable set of inter molecular attractions. Another favourable aspects of the DNA molecule is that it is a stiff polymer.

The advantage of DNA technology is the precision of DNA biosynthesis and hybridization in the process of matrix generation. The accuracy of photolithographic techniques may therefore be improved by about two orders of magnitude. Further since the diameter of DNA double size helix is only 2 nm, the lower size limit of the structures should be substantially reduced. Recently assemblies of supramolecular aggregates from DNA: derivated gold colloids use the concept of DNA as the framework of special arrangement of molecular devices. These nanocrystalline metal crystals may have actual application in laser technology. The nanocrystals were given individual "codon" sequences, by coupling with thiolated 18-mer-oligonucleotides.

Different devices are connected to a fixed codon and hence, can be sequenced in a desired order on suitable template strand-in DNA hybridization. With the help of this method even complex supramolecular structures could be constructed. DNA synthesis helps in preparing geometrical objects, from modules whose valence arms poses individual cohesive ends. Even complex structures such as polyhedron with 24 vertices can be synthesized from smaller groups. Thus utilization of DNA in biotechnological applications has other advantages as well. The process can be carried out in biological decomposition and also in aqueous phase. The DNA engineering has bright prospects in nanotechnology.

■ PROTON PUMPS

Proton pumps (Figure 8.1) are molecular structures, possessing the feature of producing continuous renewable conversion of light to chemical, mechanical or electrical energy, which could be used by nano scale objects. The process of synthesizing ATP requires energy, which is usually obtained from chemical or photochemical reactions. The proton gradients across the cell membrane, created by enzymes help ATP synthase (ATPase) synthesize ATP. When the protons circulate through these devices, they provide the means of coupling the exchange of free energy. When an ATPase reverses its role, it helps develop a proton gradient and is called a pump. It is very tough to transport protons across the membranes without assistance, because of the high energy of activation associated with the transfer of protons from polar (aqueous) to non-polar interior of the membrane. Another mechanism for transportation of protons suggests the translocation of excess protons along hydrogen bonded chains of properly aligned water molecules throughout the living cell membrane.

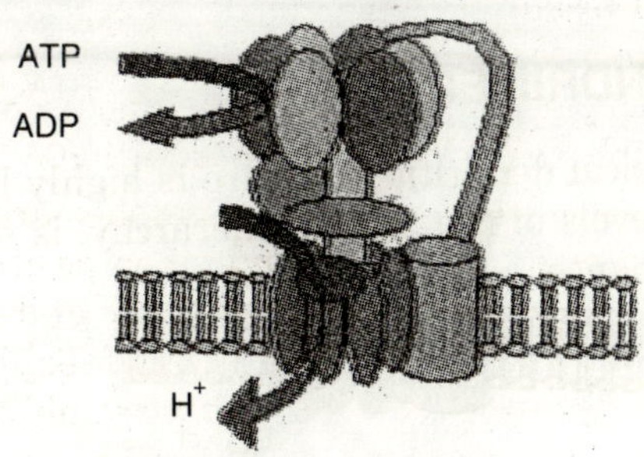

Fig. 8.1 A Proton Pump (Source : http://ntmf.mf.wau.nl/hemminga/MIVase.html)

However the proton leakages are very low because the membranes from an excellent permeability barrier. So as to achieve higher rates of proton transfer rates, assistance is provided by transmembrane proteins, providing a polar environment to the proton and reducing the activation barrier. In contrast to this mechanism, there are active transport systems such as pumps and Mitchell loops which can transport ion against concentration. Mitchell loops are redox systems which catalyze movements of electrons between reducing and oxidizing agents separated by the lipid bilayer. Pumps on the other hand are mechanisms which also are the place for energy producing chemical reactions, apart from translocation pathway.

Mere challenges lies in designing a synthetic proton pump. Although it has not yet been accomplished, several blueprints have been proposed. One of them is as shown in Figure 8.2; its details are as follows.

PS, is a proton source which initiates the proton transfer and is located near the center of bilayer of the membrane. PS transfers the proton to the primary proton acceptor A_1, which could be a part of the side chain of an amino acid in the protein. From A_1 it is transferred to A_2 and so on towards the cell membrane, until it is captured by a water molecule and is carried to the aqueous environment. The key feature required in this mechanism is to inhibit reverse proton transfer from A_1 to PS. This could be achieved by making the proton transfer between A_1 and A_2 fast and practically irreversible, by using mechanisms like conformational shifts.

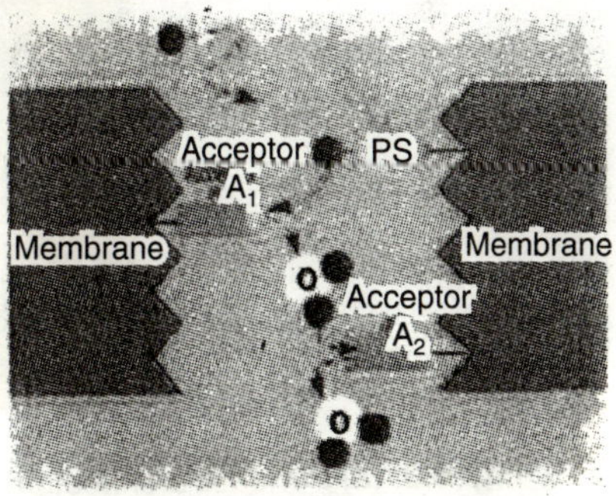

Fig. 8.2 Scheme of artificial, light driven proton pump

As far as application of proton pumps is concerned, drug delivery using pH sensitive liposome which are stable at neutral pH and release their contents on acidification. Thus with the help of controlled illumination, drugs could be released at specific location thus decreasing side effects and increasing the efficiency of the treatment. Proton pumps can also be used to regulate molecular assemblies and biochemical reactions with the control of light and pH modifications.

DENDRIMERS

The typical dendritic structure is highly branched pattern as shown in Figure 8.3, occurring at all levels of the biologic hierarchy, is endowed with the unique interfacial and functional performance.

(*a*)

(*b*)

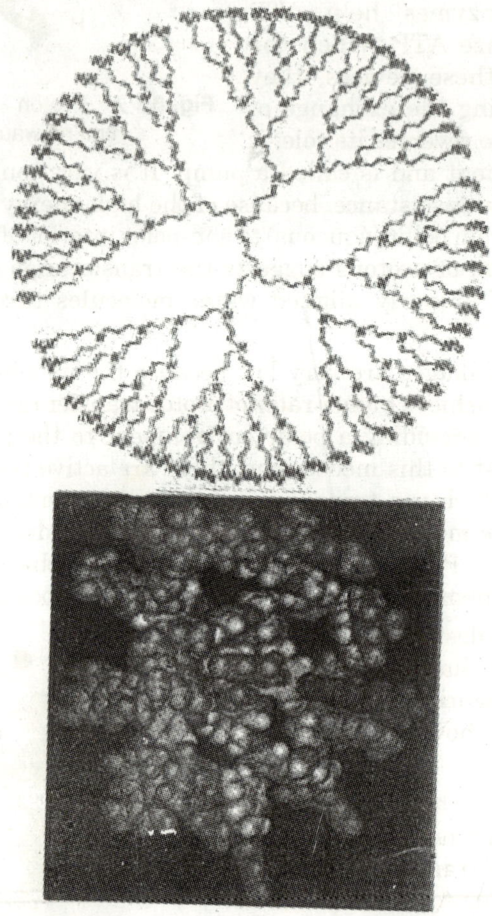

Fig. 8.3 (*a*) Dendritic architecture (*b*) a light-harvesting dendrimer performs some functions of artificial photosynthesis (source (*a*) http://www.stanford.edu/~duijvenb//dendrimers.gif; (*b*) http://www.lbl.gov/Science-Articles/Research-Review/Magazine/2001/Fall/features/02Dendrimers.html)

ally there are two strategies for dendrimer synthesis. The first one is the divergent ure 8.4*a*) in which growth of the dendron starts from the core. A second method from the molecular surface to the core. This is referred as convergent synthesis

Dendrimers can be used to capture guest molecules inside their hollow interiors. These guest molecules are unable to escape the host molecules even after extensive dialysis on surface modifications. These hollow spaces can be termed as cargo spaces which have provided the inspiration of drug delivery because entry and exit parameters can be easily controlled.

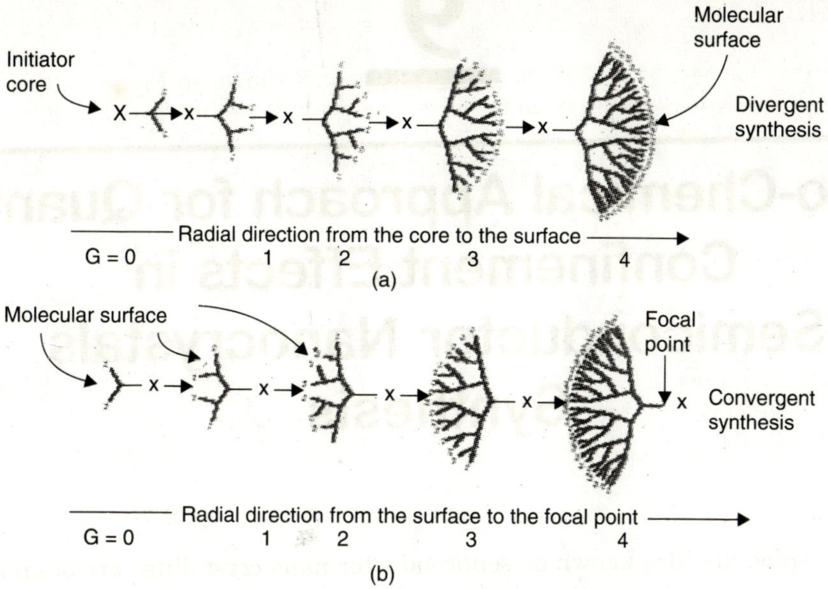

Fig. 8.4 The two principal methods of dendrimer synthesis

SURFACE MODIFICATIONS

Several biological reactions take place at surfaces and interfaces thus making surfaces play an important role in biology and medicine. Nanotechnology exploits surface and interfacial ideas and has enabled a huge variety of methods for nano fabricating and nano patterned chemistry.

Different techniques have been established for nano fabrication such as electron beam lithography, microcontact printing, self assembling structures and colloidal resists etc. The living cells respond to the imbalanced or asymmetric distribution of interfacial forces across the nano structures. Patterns of the surface chemistry possibly display binding sites for specific molecules.

The application of surface modification lies in the fact that surfaces of both high and low addition can be produced, especially in non bio-degradable ploymers. Some of the potential application would be in cell traps, bio-sensers and collagen aligners.

This branch of scientific endeavour is an interface among biology, chemistry and physics, where biology demonstrates in how to recognize, organize and functionalize including how to assemble molecular materials, which has applications in the fields like medicine to electronics. Chemical engineering and biochemical approach for biotechnological achievements should encompass a novel blend of skills to totally utilize and step for advancement in this new field.

9

Bio-Chemical Approach for Quantum Confinement Effects in Semiconductor Nanocrystals Synthesis

Quantum dots which are also known as semiconductor nano crystallites are of great interest due to their unique size dependent optical properties that render them attractive from the view point of integrated photonic devices. Semiconductor nano crystallites in a transparent matrix are of immense interest due to their potential applications in non-linear optics and optical switches. The possibility of engineering the band gap and thus tailoring the optical and electrical properties of the semiconductor leads to a great potential in materials engineering.

One among today's challenges in this frontier area of research is developing simple and effective synthesis procedures for obtaining nano structures in a form easily adaptable for device applications. The most common route for chemical synthesis involves using strong matrix supports to stabilize the particles in micelles, sol-gel glasses, vesicles and polymers. Well defined semiconductor clusters can be easily prepared by capping them with a polymer.

The signature blue shift and excionic transistions have been absorbed in optical absorption, photoacoustic and fluorescence spectroscopy.

■ PROCEDURE

The metal salt *i.e.*, lead and zinc solution is added to aqueous PVA solution in a flask and the mixture is stirred for an hour. Sulphur ion is then added to the solution that is further stirred for about three hours. The resultant brownish red solution is cast on a glass substrate and heated at 50°C for an hour. Polymer films of millimeter thickness are then peeled of from the glass plate. The samples are characterized by X-ray diffraction and transmission electron microscopy. Optical absorption spectra of the samples are recorded on a spectrophotometer, thermal analysis is done on Perkin-Elmer thermal analyzer and infrared spectra are recorded on a Bruker spectrometer.

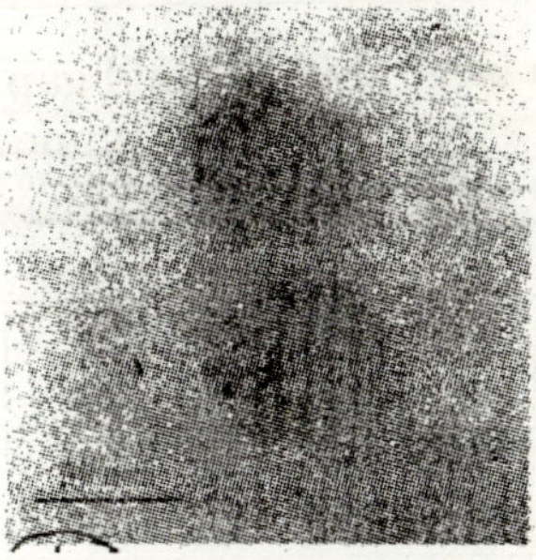

Fig. 9.1 (*a*) Transmission electron micrograph of PbS nanoclusters

Fig. 9.1 (*b*) Selected area diffraction pattern of PbS nanoclusters

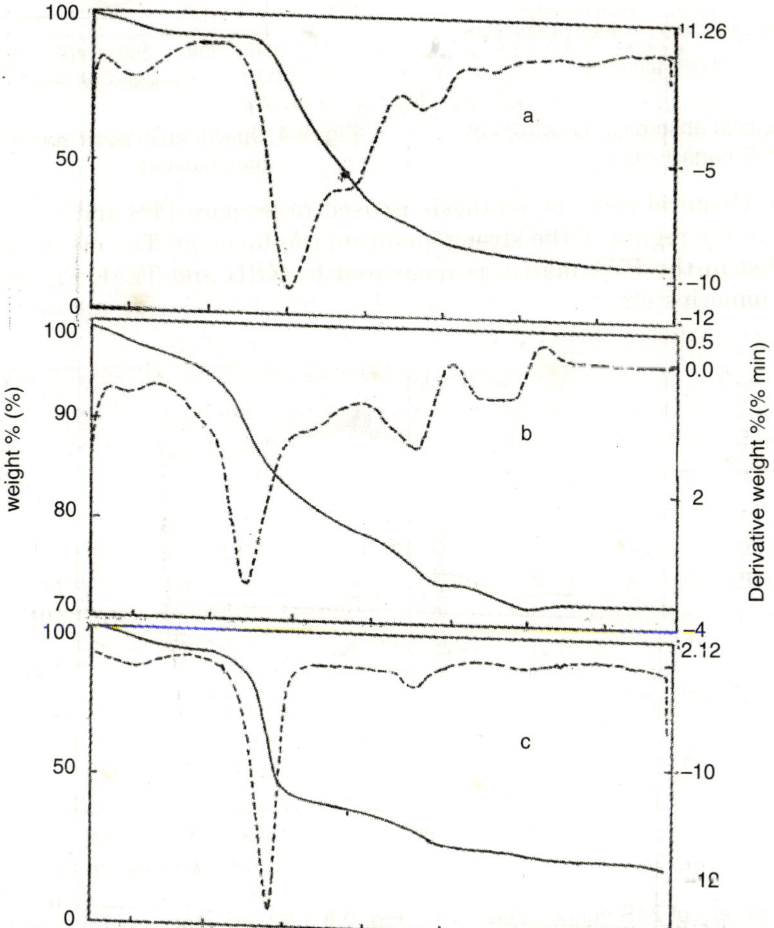

Fig. 9.2 Thermogravimeteric analysis curves for (*a*) PVA, (*b*) PbS/PVA, (*c*) ZnS/PVA

Optical absorption spectra of PbS nanoclusters show a large shift in band gap from the bulk value of 3020 nm (Figure 9.3). Such a large blue shift is expected in the regime of strong confinement, according to the effective mass approximation (EMA) model.

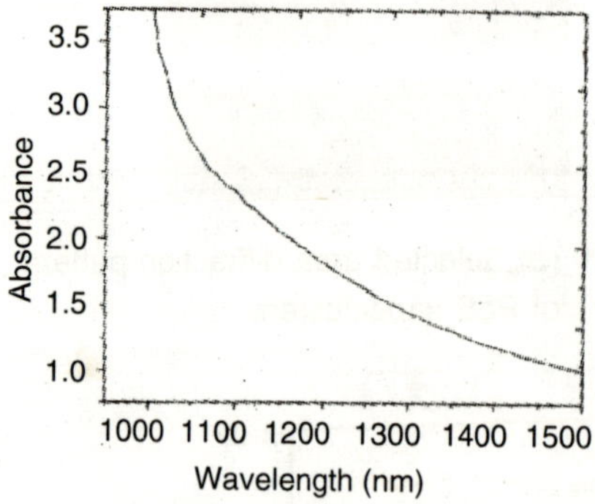

Fig. 9.3 Optical absorption spectrum of PbS nanoclusters

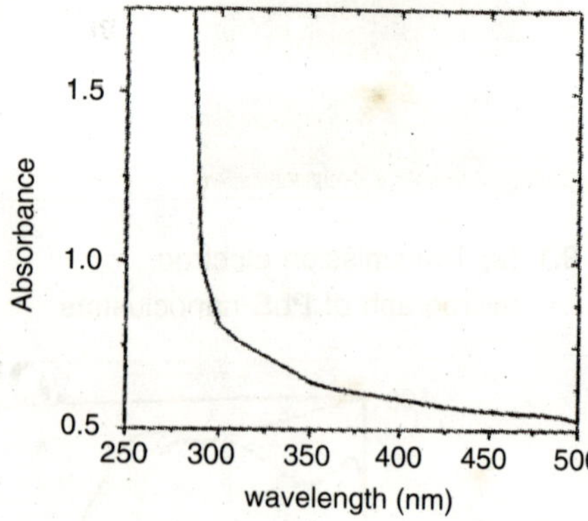

Fig. 9.4 Optical absorption spectrum of ZnS nanoclusters

A simple chemical route of synthesis is used to prepare PbS and ZnS semiconductor nano materials in the regime of the strong quantum confinement. The mean size of the nanocluster embedded in the PVA matrix is measured by XRD and TEM. Fig. 9.5 shows XRD pattern of ZnS nanoclusters.

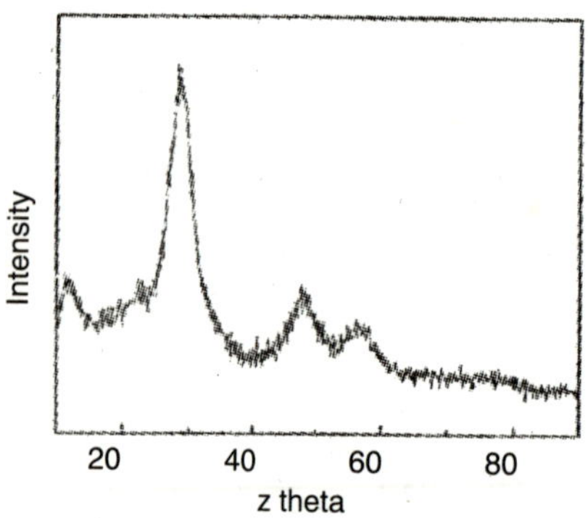

Fig. 9.5 XRD pattern of ZnS nanoclusters

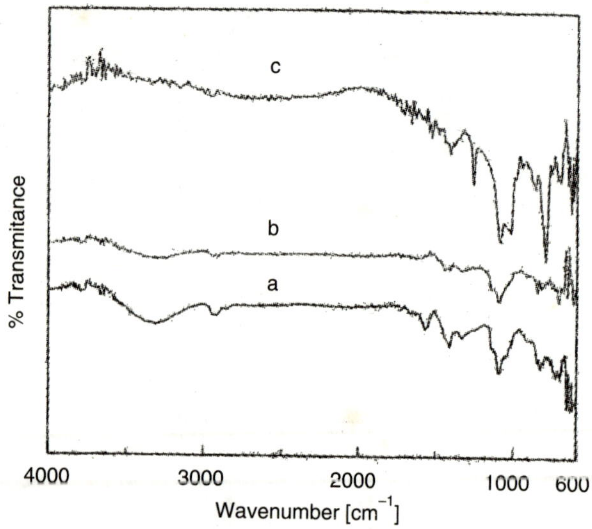

Fig. 9.6 Infra red spectra of (*a*) PVA, (*b*) ZnS/PVA, (*c*) PbS/PVA

The highly viscous polymer solution plays a key role in the preparation of nano structures. The optical adsorption spectra show significant shift of the cut-off wevelength from that of the bulk crystal, indicating the presence of nano clusters in the regime of strong quantum confinement. Result of thermoagravimetric measurements and infrared spectroscopy Fig. 9.6 indicate the role of the polymer host matrix to be purely physical in nature.

10

Light and Heat Induced Metastability in Nano Crystalline Silicon

Porous silicon (PS), is a form of nano crystalline silicon, and has attracted more attention due to the possibility of its wide range of applications in optoelectronic devices, vapour sensors, gas sensors and chemical sensors. PS consist of nanometer size crystallites of silicon and shows photoluminescence (PL) which is visible under room temperature. PS surfaces are naturally terminated with a covalently bonded monolayer of hydrogen but this monolayer does not give protection against chemical absorption and exposure to light. Further it has been found that after a short light exposures the dark current (DC) is higher then the annealed state dark current. This higher dark current is called persistent photocurrent (PPC), persists for several hours at room temperature. It has been reported that the LS decreases PL and increases electron spin resonance (ESR).

It has been reported that the dangling bond density in the nano crystalline silicon decreases after light exposures of short durations. This is accompanied by an increase in DC, photocurrent (PC) and PL. Rapid cooling from high temperature to room temperature also gives an increase in DC, PC and PL and a decrease in ESR signal. Such an increase in DC after fast cooling has been absorbed by various scientists in doped hydrogenated amorphous silicon also. It is also suggested that the metastable changes caused by the thermal quenching and light exposures might be related to the structural changes in PS.

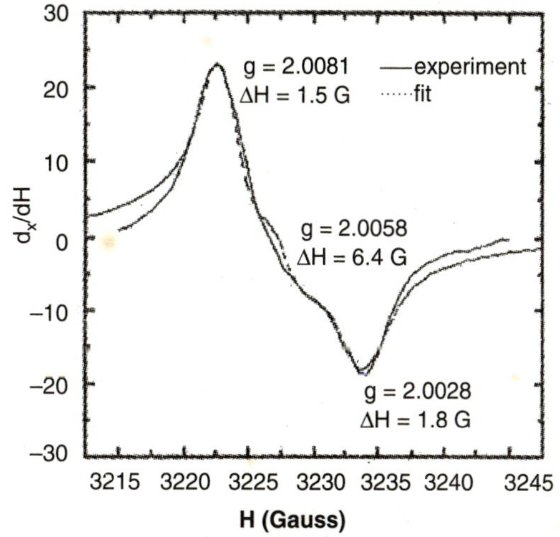

Fig. 10.1 ESR signal in state A. Dark line is experimental and dotted line is the fitted data

Figure 10.1 shows the ESR signal of PS in the annealed state. The ESR signal is asymmetric and is similar to that reported by others. This can be fitted to three signals; one isotropic signal with $g = 2.005 \pm 0.0006$

and $g = 2.0081 \pm 0.0006$ and $g = 2.0028 \pm 0.0006$. The last two readings have been attributed to the dangling bonds in PS by some nanotechnology scientist.

The decrease in ESR signal accompanies an increase in PC and PL from short exposures. This indicates that decrease in dangling bonds, which probably act as the nonradiative recombination centers, is the cause of the rise in the PC and PL. However, long exposures have opposite effect. It is also found that for long exposures the dangling bond density (ESR) increases and PL decreases. However short exposures do not have any effect on PL whereas ESR decreases.

It is also observed that an increase in DC, PC, PL when PS is exposed to light for a short duration (< 2700 s). This is accompanied by a decrease in ESR signal. For exposures longer than 3000 s we get opposite effect. Fast cooling of PS to room temperature from 150°C gives a state with higher DC, PC and PL, and is also accompanied by a decrease in the ESR signal. Further, all metastable states can be annealed out at 150°C (1 h), but are stable against exposure to sub-band gap (IR) light. This indicates that the observed changes cannot be explained by trapping of charges alone. In this connection it is suggested that a local structural rearrangement of bonds involving movements of hydrogen atoms is responsible. This movements results in decrease in dangling bonds for the short exposures and for the fast cooling experiments. This can explain the observed decrease in ESR and the increase in PC and PL for short exposures and for fast cooling. Further, the presence of nano crystalline silicon also seems to play a role in these rearrangements since no such decrease in ESR has been reported in a—Si: H, so far. On the other hand for long exposures, the breaking of weak Si—Si seems to be dominant effect which gives an increase in ESR resulting in decrease in DC, PC and PL.